The Radio Today guide to the Yaesu FTDX10

By Andrew Barron ZL3DW

This book includes useful tips and tricks for the configuration and operation of the fabulous Yaesu FTDX10 transceiver. Rather than duplicate the manuals which describe each button, function, and control, I have used a more functional approach. This is a "how to do it" book with easy-to-follow step-by-step instructions.

The author has no association with Yaesu, any Yaesu reseller, or Yaesu service center. The book is not authorized or endorsed by Yaesu or by any authorized Yaesu dealer or repair center. Research material for the creation of this document has been sourced from a variety of public domain Internet sites and information published by Yaesu including the FTDX10 Operation Manual. The author accepts no responsibility for the accuracy of any information presented herein. It is to the best of my knowledge accurate, but no guarantee is given or implied. Use the information contained in this book at your own risk. Errors and Omissions Excepted.

Radio Today is a trademark of the Radio Society of Great Britain www.rsgb.org and is used here with their permission.

Cover graphics by Kevin Williams M6CYB. The transceiver image and Yaesu logo are used with permission from Yaesu (UK) Ltd.

Yaesu, Yaesu Musen, and the Yaesu logo are registered trademarks of Yaesu Musen Co, Ltd. (Tokyo Japan). Yaesu USA, Yaesu UK, are registered Yaesu companies.

Microsoft and Windows are registered trademarks of Microsoft Corporation.

All other products or brands mentioned in the book are registered trademarks or trademarks of their respective holders.

The Radio Today guide to the Yaesu FTDX10

Table of Contents

OTHER BOOKS BY ANDREW BARRON

Testing 123
Measuring amateur radio performance on a budget

Amsats and Hamsats
Amateur radio and other small satellites

Software Defined Radio
for Amateur Radio operators and Shortwave Listeners

An introduction to HF Software Defined Radio
(out of print)

The Radio Today guide to the Yaesu FTDX101

The Radio Today guide to the Icom IC-705

The Radio Today guide to the Icom IC-7300

The Radio Today guide to the Icom IC-7610

The Radio Today guide to the Icom IC-9700

ACKNOWLEDGEMENTS

Thanks to my wife Carol for her love and support and to my sons James and Alexander for their support and their insight into this modern world. Thanks also to Yaesu who produced the excellent FTDX10 transceiver and finally, many thanks to you, for taking a chance and buying my book.

ACRONYMS

The amateur radio world is chock full of commonly used acronyms and TLAs (three letter abbreviations :-). They can be very confusing and frustrating for newcomers. I have tried to expand out any unfamiliar acronyms and abbreviations the first time that they are used. I have assumed that anyone buying a radio such as the FTDX10 will be familiar with commonly used radio terms such as VFO, AGC, SSB, CW, MHz, and kHz. Near the end of the book, I have included a comprehensive glossary, which explains many of the terms used throughout the book. My apologies if I have missed any.

Approach

Rather than duplicate the Yaesu manual which describes each button, function, and control, I have used a more functional approach. This is a "how to do it" book. For example, I describe how to set up the transceiver for SSB operation. Then I follow that up for CW, FM, RTTY, PSK, and external digital mode software such as FT8.

Some people complain that this book is a "rehash" of the manual. I would like to point out that this book is a user guide to the FTDX10. The Yaesu manual is also a user guide to the FTDX10. Expect some similarities! If I left out some functions or controls, I am sure people would complain about a lack of completeness. The aim is not to replace the manual but to more fully explain how to configure and operate the radio to take advantage of its many great features. For example, when I cover the front panel controls, I explain not just what the control does, but how and when to use it. Some of the buttons are disabled in some situations. Along the way, I offer a few 'tips' on how I configured my radio. You don't have to follow these suggestions, but they provide some guidance.

The FTDX10 requires some initial configuration, especially if you want to use external digital mode software such as WSJT-X (FT8), Fldigi, MixW, CW Skimmer, or MRP40. There is a range of color options for the spectrum display. I imagine that you will experiment with some settings more than once before you decide on the optimum settings for your radio.

I cover updating the radio firmware, loading the Windows driver software for the USB cable connection, and connecting the radio to your PC for CAT control. I even cover the FM mode, how to set up tones for repeater access, and store repeater channels in the radio memory slots.

The 'Setting up the radio' chapter is followed by information about 'Operating the radio' in various modes including using Split, Contour, APF, and the various noise reduction and notch filters. Then I cover some ways of operating the radio including programming and using the voice keyer and the CW, PSK, and RTTY message keyers, and a few tips on using the radio for the FT8 mode.

The Troubleshooting section deals with some oddities that might trip you up. The chapters about the front panel controls and the front and rear panel connectors, supplement the information provided in the Yaesu manuals. Adding more information about what each control does and how to use them. Finally, there are chapters describing each of the Soft Key functions and the FUNC menu settings.

The Glossary explains the meaning of the many acronyms and abbreviations used throughout the book and the Index is a great way of getting directly to the topic you are looking for. Right at the end, there is a handy quick reference guide, listing most of the menu commands in alphabetical order.

Conventions

The following conventions are used throughout the book.

Front panel knobs and buttons are indicated with a highlight.   MODE

Touchscreen controls are indicated without a highlight.   RF POWER

'Touch' means to briefly and gently touch the item on the touchscreen. There is no need to press hard on the screen. Most touchscreen controls are indicated with a beep as well as changing the function or opening a window or menu.

'Touch and hold' means to keep touching the screen icon until a secondary function or popup window is displayed. This is usually accompanied by a double beep.

'Press,' 'Push,' or 'Select' means to press a physical button or knob. The action is usually indicated with a beep as well as changing the function or opening a window or menu.

'Press and hold' or 'hold down' means to hold a physical button or control down for one second until the function changes. It usually opens a control option or sub-menu window on the touchscreen. Press and hold actions are usually accompanied by a double beep.

A 'Soft Key' is an icon on the touchscreen that represents a button or control.

Words in <brackets> indicate a step in a sequence of commands. The sequence usually begins with pressing a <button> or a <knob> followed by touching a series of <Soft Keys> or <icons> on the touchscreen.

FUNC means to press the function knob on the front panel.

Select <FUNC> means, press the FUNC knob, and select the menu option by touching the screen. Or you can select the menu option by turning and clicking the FUNC knob. To change the value, you can usually turn the VFO or FUNC knob, or use the left < and right > arrow icons. A typical sequence might be, <FUNC> <Display Setting> <Display> <My Call> <*enter text*> <ENT> <FUNC>.

MODE means to press the mode button on the front panel or touch the mode icon.

'Mode' usually means one of the transceiver modulation modes. SSB, CW, RTTY, PSK, AM, FM, or Data. But it can also mean the 'CENTER', 'CURSOR' or 'FIX' scope display mode, or an external digital mode performed by software running on your PC, such as FT8 or PSK31.

The Yaesu FTDX10

Congratulations on buying or being about to purchase the fabulous Yaesu FTDX10 transceiver. It is always exciting unboxing and learning how to use a new transceiver. Like many of you, I was expecting the FTDX10 to be a "cut down" version of the highly successful FTDX101D. But while it does share a lot of the same technology, the FTDX10 has several features that are not available on the FTDX101, and some of its performance figures are slightly better. The radio has superb technical specifications, making it an elite performer for a budget price. Currently, the FTDX10 is number three on the renowned Sherwood Engineering performance table, http://sherweng.com/table.html.

The FTDX10 has similar styling to the FTDX101, and it is easy to see that it is from the same Yaesu family. The front panel is dominated by a bright full-color 5" touchscreen display and the VFO tuning knob which has an outer multi-function tuning ring surrounded by the buttons that enable the functions it controls. There are "proper" buttons and knobs for often-used functions such as Noise Blanker, Split, and VFO switching. Other functions such as the band and mode switches, direct entry of a frequency, and adjusting the transmitter output power have been moved on to touchscreen controls.

Under the skin, the radio employs the same 'Hybrid SDR' (software defined radio) receiver architecture as the FTDX101. The receiver has a superheterodyne stage, down converting the incoming radio spectrum to a 9 MHz I.F. Then a narrowband SDR 'receives' the I.F. bandwidth and converts it down to a 24 kHz second I.F. for the DSP (digital signal processing) stage. Unlike recent releases from other manufacturers, the radio is not a direct sampling software defined radio. Yaesu has opted for the hybrid technology to ensure top of the range IMD (intermodulation distortion) performance. The radio does include a 'direct sampling SDR' receiver but it is only used for the Band Scope display.

If this radio is your first SDR, I am sure that you will be excited about the spectrum and waterfall display. Yaesu doesn't call it a band scope or a panadapter, simply referring to it as the 'Scope' or 'Display.' The graphics are similar to the FTDX101 display, but the screen is crisper with a higher resolution, thanks to the smaller screen, and in my opinion, the onscreen graphics are much better. You can change the brightness of the TFT display and the LED indicators.

The FTDX10 is a truly exceptional radio for contesting or working DX stations, with a fantastic receiver. The spectrum scope can display a band of frequencies up to 1 MHz wide. More than enough to show contest stations across any of the lower bands, or the pileup waiting to work a DXpedition.

I love the SSB voice message keyer. It makes calling "CQ Contest" or just saying my callsign repeatedly, so much easier.

You can operate CW, PSK, or RTTY without relying on a connection to digital mode software running on a PC. There are built-in decoders for all three modes. But transmitting is limited to sending five pre-set messages (macros). Although you can use an external USB keyboard to pre-load the text messages, you cannot use it to send text directly from the keyboard. The CW keyer can send a contest number that increments each time the text macro is sent. Sadly, this feature was not extended to the PSK or RTTY modes. The inability to send free text severely limits the usefulness of the internal PSK and RTTY decoders and I expect that most digital mode operators will continue to use external digital mode software. However, you could work a DXpedition or operate in a CW contest using the built-in software. Especially if you were only using the CW decoder to confirm callsigns and contest exchange numbers.

Mouse operation exactly duplicates touching the screen with your finger. The right mouse button and center wheel are not implemented, which is a pity. The right mouse click could have opened applicable menus and the wheel could have been used for fine-tuning. The Logitech M310 and M305 wireless models will work, and probably some other mice, or you can use a 'wired' mouse that has a USB cable.

The 'Filter Function Display' is great. It shows the audio bandwidth including the action of the manual notch filter or contour control. The noise reduction system and noise blanker are excellent.

The FTDX10 has a function screen accessed by pressing the FUNC knob. The screen displays a bewildering array of 26 options plus five 'Soft Keys' that access deeper menu layers. In most cases, turning then pressing the FUNC knob is used to make menu selections, or you can just touch the relevant icon on the screen.

I cover some "weird" operating behaviors in the troubleshooting chapter. If you experience something strange, have a look to see if it is covered in that chapter. I am sure that some of these oddities will be addressed in future firmware releases.

The FTDX10 is a 'contest grade' transceiver with a fantastic receiver and virtually every option that you could want. It combines cutting-edge SDR technology with all the controls and features that experienced amateur radio operators expect from a Yaesu transceiver. Setting up the radio is relatively easy. But there are a lot of options that you can apply. I will cover most of them as we go through the 'step by step' instructions for setting up the radio and discussions about all the functions and features. I included an alphabetically sorted 'quick reference guide' at the back of the book to make it easier to find your way through the menus and make the changes that you want. There is a lot to learn. So, let's get started.

TECHNICAL FEATURES

The FTDX10 is a hybrid SDR. The receiver uses a superheterodyne stage to convert the incoming signals down to a 9 MHz I.F. Followed by an SDR stage which converts the signal down to a digital 24 kHz DSP stage. That is followed by the audio amplifier. The transmitter is also an SDR design with the final frequency being generated directly by a 16-bit DAC with no up-conversion mixer.

At the start of the receiver chain, there is a 6 dB attenuator followed by a 12 dB attenuator. They can be enabled using the ATT Soft Key on the display to provide 6, 12, or 18 dB of front-end attenuation. The attenuators are simple voltage dividers consisting of two resistors each.

They are followed by 15 automatically selected bandpass filters. Rather more than in most transceivers. Ten of them are dedicated to amateur bands and the remaining five are for the general coverage receiver. The 10 dB AMP1 RF preamplifier is always in the circuit. You can switch in AMP2, before AMP1. Or you can switch in the IPO (Intercept Point Optimization). IPO introduces about 10 dB of attenuation and AMP2 introduces around 8-10 dB of gain. The three options are switched in or out with the IPO Soft Key. The 'normal' setting is to use AMP1.

At this point, some of the wideband signal is split off to feed the 'direct sampling SDR' receiver, which is used solely for the spectrum scope and waterfall display.

The RF amplifier is followed by the first mixer and local oscillator stage to create the 9.005 MHz I.F. passband. An SDR style ADC (analog to digital converter) and FPGA (field programmable gate array) converts the I.F. passband into a digital signal, which is passed on to a 32-bit Texas Instruments TMS320C6746 DSP (digital signal processing) stage for filtering and demodulation. Finally, a DAC (digital to analog converter) recovers the audio output signal for the audio line output, speaker, and headphones. The Texas Instruments DSP chip provides the I.F shift and width, controls, notch and contour filters, APF (audio peak filter) for CW, the noise blanker, and the digital noise reduction algorithms.

The outstanding 116 dB RMDR (reciprocal mixing dynamic range), 141 dB BDR (blocking dynamic range), and 109 dB 3rd order IMDR (intermodulation distortion dynamic range) figures are the result of Yaesu using an exceptional local oscillator and double balanced mixer for the first stage of the receiver, and the high-performance crystal roofing filters. By the way, the basic MDS (minimum discernible signal) sensitivity and BDR figures are slightly better than an FTDX101D measured by the same lab. The local oscillator is a 0.5 ppm high stability TCXO (temperature-controlled crystal oscillator) driving a 'high resolution direct digital synthesizer' operating at 250 MHz. The oscillator output is divided down to the required local oscillator frequency, to produce a local oscillator with an outstanding phase noise performance, better than -145 dBc/Hz at a 2 kHz offset.

The receiver's selectivity and 3rd order IMD dynamic range is greatly improved by the sharp narrowband roofing filters working at the 9 MHz I.F. The FTDX10 has 500 Hz, 3 kHz, and 12 kHz roofing filters fitted as standard. A 300 Hz filter can be added as a user-installed option. There are no variable DSP RF filter options because the radio is not a direct sampling SDR. But you can use the IF width and shift controls.

The transmitter is also an SDR design with the final frequency being generated directly by a 16-bit DAC with no up-conversion mixer. The 250 MHz 'high resolution direct digital synthesizer' used in the receiver also contributes to the outstanding phase noise characteristic of the transmitter. A clock distributor divides and distributes the clock signals to the transmit blocks including the FPGA and DAC. Transmitter phase noise is an impressive -145 dBc/Hz at a 2 kHz offset. The FTDX10 that the ARRL tested for their QST review had slightly better 'worst-case transmitter IMD' and a narrower CW keying spectrum than the FTDX101D.

If you have not used an SDR transceiver before, you will be impressed by how clean the receiver sounds and you will quickly get used to the advantages of the spectrum and waterfall display. The layout is different from previous Yaesu radios, but anyone transitioning from other Yaesu radios will be comfortable with the controls.

The touchscreen is excellent, I like it a lot more than the touchscreen on the FTDX101D. But the spectrum scope is not as configurable as the scopes on competitor's radios. There is a limited choice of colors, you cannot change the dynamic range, and there is no averaging function. The smaller screen on the FTDX10 is crisper than the one on the FTDX101 with the same 800x480 pixel resolution. It has about 185 dots per inch, which is roughly 25% more than the FTDX101. I like the graphics on the FTDX10. Everything looks smart and clear, with a good choice of color highlights. A few items have been renamed to clarify their function. For example, the 'REC/PLAY' Soft Key used to turn on the message keyers has been changed to 'MESSAGE.' The menus look better, and if you hold an arrow down, the screen will scroll through the menu list. Unfortunately, the FIX spectrum display is very badly implemented. You should be able to select any start and stop frequency with a span of less than 1 MHz. Other transceivers allow four fixed spectrum display ranges per band. And there should be a dedicated control for the spectrum scope level. Or it should auto-level between bands. Luckily, LEVEL is one of the controls that can be dedicated to the large tuning ring around the VFO knob. Surprisingly, this is a good place for this setting. I leave it allocated that way most of the time.

There is no beep or on-screen indication when you tune outside of a ham band. The only change is a faint click as the bandpass filter switches from the ham band to general coverage at the end of a 500 kHz band segment.

The 3DSS ('three dimensions signal stream') is a spectrum display that includes signal history, similar to the information on a waterfall display. It is represented as a "3D" image tapering back to give an illusion of depth. Yaesu is very excited about this way of representing received signals, but I prefer the traditional spectrum and waterfall display. The 3DSS display looks very flash when the radio is on the display stand at a ham convention or trade show, but I don't believe that it adds much value when you use the radio at home.

The IPO (intercept point optimization) mode is designed to improve the receiver's dynamic range and intermodulation performance. This does come at the expense of a 10 dB reduction in receiver sensitivity. Selecting IPO will improve the signal to noise ratio, especially on the noisy low bands. In the event of a very strong signal causing intermodulation problems, you are recommended to try using IPO first, then add in the front-end attenuators if the offending signal is still too strong. The 'normal' operating condition is to select AMP1. Turning on AMP2 adds a second amplifier to provide an additional 8-10 dB of gain. You might use AMP2 on the 10m and 6m bands.

The radio has a three-stage parametric equalizer for tailoring the transmitted signal to your voice and microphone. Unlike the FTDX101, the FTDX10 also has a three-stage audio equalizer for the received audio.

Another feature that the FTDX10 has, that the FTDX101 does not, is the ability to record signals 'off-the-air' and save them onto the SD memory card. You cannot transmit the recordings, but you could copy them to your PC and transmit them from there.

I find the 'Contour' control very useful. It creates a shallow manually tuned notch filter which can be used in that way to combat weak interfering signals. Or, if you want to use it as a tone control, you can adjust the notch frequency to cut or boost either the high or low audio frequencies. The contour notch frequency is indicated by an orange dip or bump on the filter function display. There are menu settings to adjust the width and depth of the notch or boost it provides. The effect of the filter is easy to hear, and it can be seen on the filter function display, or the AF-FFT display if the audio scope displays have been enabled using the MULTI Soft Key.

I like the APF (audio peak filter) for CW and the ZIN button which pulls CW signals to the correct tone. There is also a tuning indicator for CW under the S meter.

Some online commentators say that they don't like touchscreens. This is NOT a radio for them. It is impossible to use the radio without touching the screen. All of the menu structure and even basic functions like changing the output power can only be changed via the touchscreen display. Anyway, I really like it. It works very well, with good sensitivity and it is crisp, bright, and colorful with high definition.

Setting up the radio

The Yaesu manual does a good job of identifying all the controls and menu items, but it lacks detail in some areas. In this chapter, I cover the processes that you need to follow to get ready for operating the radio. Each section lists the process that you need to follow when you set the audio levels, tone settings for FM, and transmitter bandwidth. It is important to set the Mic Gain, AMC (automatic microphone gain), and compression controls correctly for SSB operation. I explain the setup for the FM, CW, RTTY, and PSK modes including the built-in options and using external digital mode programs. There are instructions on configuring the five voice messages for SSB and the keying messages for CW, RTTY, and PSK.

I tell you how to download and install the USB driver software to allow the radio to communicate with your computer and discuss setting up the virtual COM ports and the Audio CODEC, linear amplifier connections, setting up the Presets, and the incredibly popular FT8 mode. Finally, I cover formatting and using the SD card, how to do firmware updates, and how to set the clock.

The radio is easy to configure once you know which settings to use. Where I have changed my radio from the Yaesu default settings, I have explained why I made the change and the effect that it has on the radio. The big advantage of using the instructions in this section rather than using the Yaesu manual is that I have included all the necessary steps, and the optional ones, in one place. I tell you what the controls and settings do and how they should be adjusted.

DISPLAY YOUR CALLSIGN WHEN THE RADIO STARTS

It is nice to personalize the radio by having it display your name or callsign as the radio 'boots up.'

Press <FUNC> <DISPLAY SETTING> <MY CALL> and use the onscreen keyboard to change the message. You are allowed 12 characters. There was just enough room for me to put 'ZL3DW FTDX10.' Press ENT when you have finished, to save the changes, then BACK, BACK, or FUNC to exit.

SETTING THE TIME AND DATE

The radio has a clock, but it is never displayed on the screen. It is only used for time stamping the filename of saved files. Press <FUNC> <EXTENSION SETTING> <DATE&TIME> and set the DAY, MONTH, YEAR. Press <FUNC> to save the date. Then set the HOUR, & MINUTE and press <FUNC> to save the time.

Tip: If you want your files date stamped with UTC date and time, you can enter UTC date and time details. But I believe it is easier to use local time.

GETTING READY FOR SSB OPERATION

Three settings should be adjusted before you start transmitting on SSB. You need to set the Microphone Gain control, the AMC (automatic microphone gain) control, and the Speech Processor (PROC) level. AMC is an automatic leveling control that prevents you from overmodulating the transceiver. You may not intend to use the speech processor (compressor), but you might as well learn what PROC LEVEL setting you need to set it to if you do decide to use it.

The AMC and Mic Gain settings are very broad, and it is difficult to get a satisfactory result. You might prefer to just set the three parameters to the settings in the 'SSB Audio Settings' table below and forget about the adjustment procedure. But if you want to "give it a go," carry on with the steps below.

TIP: These settings also apply to the AM and FM modes, so there is no additional setup required for AM or FM. Although you can set the microphone level on those modes using the<FUNC> <RADIO SETTING> <mode> <MIC GAIN> menu if you want to.

If you want to listen to the transmitted audio to check the quality, or just hear "how it sounds," you can turn on the audio monitor by pressing <FUNC> <MONI LEVEL> and then adjusting the monitor volume with the FUNC knob. For voice modes such as SSB, it is best to listen to the monitor through headphones.

Typical test setup

It is always preferable to make transmitter adjustments with the radio connected to a suitable high power (100W) 50 ohm load. A high power (100W) 50 ohm, 50 dB, or 60 dB, attenuator will work just as well.

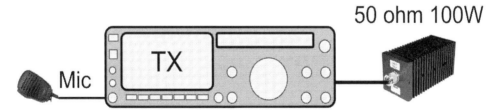

If you do not have a suitable load, you can set up the SSB levels while transmitting into your normal antenna. Preferably carry out the adjustment during the daytime when the HF bands are closed and announce your callsign regularly. You can set the SSB levels on any HF band, provided you have a suitable RF load or antenna.

Note about the AMC Level

Some Yaesu FTDX101 forum members are saying that you **must** use an AMC Level below 55 and that the radio will produce *"lots of products/whiskers/splatter"* if you don't. This is complete rubbish. I have tested the AMC leveler performance, 2 tone IMD, envelope distortion, and the RF spectrum during speech modulation.

There is no trace of RF splatter at any AMC setting, nor would I expect there to be. I assume that they have failed to calibrate the spectrum scope for transmitting and are misinterpreting the spectrum scope display. See page 180.

SSB Setup

1. Set the radio to USB. Or LSB if you are performing the setup procedure on one of the bands below 10 MHz. I usually use 14.200 MHz.

2. It is important to set the following initial conditions before you begin. If the AMC LEVEL is too low, or too high, you will not be able to set the Mic gain properly.

    a. Press <FUNC> <AMC LEVEL> and use the FUNC knob to set the AMC level to 60.

    b. Press <FUNC> <PROC LEVEL> and use the FUNC knob to set the speech processor PROC LEVEL to off.

    c. Press <FUNC> and check that the MIC EQ is off. (You can set that up later).

    d. Press <FUNC> <MIC GAIN> and use the FUNC knob to set the Mic Gain to 30.

3. It is a personal preference, but I set the AMC RELEASE TIME to the FAST setting during the setup process. It reduces the time that the COMP meter takes to settle after each voice peak. If you want to, you can set it back to the default MEDIUM setting after you have finished adjusting the transmitter audio levels. <FUNC> <OPERATION SETTING> <TX AUDIO> <AMC RELEASE TIME> <FAST>.

4. Double check that the radio will be transmitting into a suitable 100 watt (or higher), 50 ohm RF load, or a resonant antenna.

Method

1. Touch the meter display and set the meter to display ALC.

2. Check the indicator under the VFO numbers says, 'MIC GAIN.' If it does not, press <FUNC> <MIC GAIN>.

3. Key the transmitter with the microphone PTT switch and talk in your normal 'on-air' voice. In all the excitement is tempting to talk louder or shout. But you want to modulate the transmitter at the same level you will be using when you are using the radio. I usually say my callsign and "testing 12345 54321" a few times. You could use a typical CQ format, but someone might come back to your call. Watch the ALC meter and adjust the MIC GAIN using the FUNC knob, so that the voice peaks are reaching the top area of the white ALC range and not peaking into the blue range.

Don't worry if the average peaks are not quite reaching the top of the white scale. A whistle should push the meter into the blue area. The meter is very damped, but normal speech should fall mostly in the white zone.

4. Don't make big adjustments to the microphone level. It is better to adjust the MIC GAIN and then the AMC LEVEL in small increments until you are happy with both the ALC metering and the COMP metering.

5. The microphone level may be lower than you were expecting. The Yaesu default level is 30 and I found that 35 was too high. Much lower than 30 and the ALC meter will not reach the top of the white ALC range. Much higher and it will be peaking well into the blue zone.

    TIP: The level of the microphone gain is not particularly critical. If it is set too high the AMC will work harder to make sure that the transceiver never gets overloaded. I did not see any clipping distortion at any setting.

6. Now touch the meter and select COMP. Press <FUNC> <AMC LEVEL> and set the AMC level with the FUNC control. With the AMC LEVEL set between 60 and 80, the COMP meter should spend most of the time between 5 and 15 dB, with peaks up to 20 dB or higher.

    TIP: The Yaesu operating manual says that AMC should be adjusted so that the COMP meter never deflects past 10 dB. This seems to be impossible and is causing all sorts of confusion on the forums. My advice is to ignore the statement in the manual and make an adjustment that swings over most of the meter scale rather than sitting mostly in the high region. A setting of 60 seems appropriate.

➢ **Setting the speech processor compression level**

The FTDX10 is a bit unusual in that you have to set the compressor level every time you use the speech processor. It is not such a chore if you remember what you set the control to last time. But still, it is a bit odd. Even if you don't plan to use the speech processor, I suggest you carry out this procedure so that you know the correct setting, in case you do want to use it in the future.

Setup

This adjustment carries on from the previous adjustments. The setup is identical.

Method

1. Leave the meter set to COMP. Press <FUNC> <PROC LEVEL> to allocate PROC LEVEL to the FUNC control. The speech processor is indicated with a PROC icon below the COMP meter.

2. Transmit into a suitable RF load or resonant antenna, key the transmitter with the microphone PTT switch and talk in your normal 'on-air' voice.

3. While observing the compression indicated on the COMP meter, use the FUNC knob to adjust the PROC LEVEL. Yaesu recommends that the compression level be kept under 10 dB. I don't like to use a lot of speech compression because it can make your transmission sound strange. So, I have my PROC LEVEL adjusted for normal speech to be peaking between 5 dB and 10 dB. I ended up with the PROC LEVEL set to 12.

The MIC GAIN and AMC LEVEL settings are difficult to get right, and they are not super critical. The following settings gave me a result that is close to the meter readings indicated in the Yaesu manual.

SSB AUDIO SETTINGS		
Function	ZL3DW Setting	My Setting
MIC GAIN	30	
AMC LEVEL	60	
PROC LEVEL	12	

➢ **The speech processor**

These pictures show the effect of the speech processor on the transmitted power. The average power is increased but the peak power remains the same.

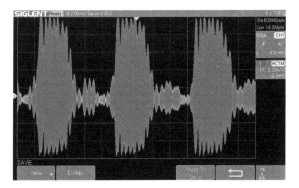

Figure 1: SSB without speech compression.

Yaesu FTDX10 on the 20m band. Test by the author.

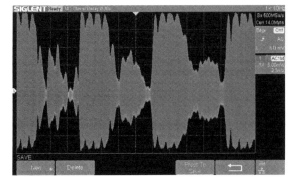

Figure 2: SSB with compression set to 15.

Yaesu FTDX10 on the 20m band. Test by the author.

SETTING THE TRANSMITTED BANDWIDTH

You can set the bandwidth of the transmitted signal for the SSB, AM, or PSK/DATA modes.

Press <FUNC> <RADIO SETTING> <MODE SSB> <TX BPF SEL>

Press <FUNC> <RADIO SETTING> <MODE AM> <TX BPF SEL>

Press <FUNC> <RADIO SETTING> <MODE PSK/DATA> <TX BPF SEL>

The default for SSB is 100-2900 Hz (2.8 kHz BW), but I set my radio to 200-2800 Hz (2.6 kHz BW) because I like to work contests and DX.

I left my radio at the default 50-3050 Hz (3 kHz BW) option for AM because it seems reasonable, and I never use AM anyway.

I changed the PSK/Data mode bandwidth to 100-2900 Hz (2.8 kHz BW) because it almost matches the bandwidth of the 200-3000 Hz receiver passband displayed on my digital mode software. The default 300-2750 Hz (2.4 kHz BW) seems a bit restricted.

AUDIO EQUALIZER (RECEIVE)

The 'FTDX10_Catalog_EN' glossy brochure includes details about a "3-stage Parametric Equalizer" for the receiver audio. It says, "*The 3-stage Parametric Equalizer tunes the low/middle/high frequencies of the received audio and is tailored for each AM / SSB / FM mode.*" However, the Yaesu manual does not mention the receiver 'equalizer' at all. It turns out that it does exist, which is great because the FTDX101 lacks any kind of receiver tone controls. Score one for the FTDX10!

➢ **For the CW mode**

You can apply up to 10 dB of boost or a 20 dB cut to the bass, mid-range, and treble. I recommend leaving the AF MIDDLE TONE GAIN at zero and applying a little boost or attenuation to the AF BASS GAIN and/or AF TREBLE GAIN as you see fit. Boosting or cutting all three should be avoided as it will cause a lumpy audio passband.

Select <FUNC> <CW SETTING> <MODE CW> and turn the FUNC knob or use the < > keys to adjust the value of each of the three, tone controls.

➢ **For all other modes**

You can apply up to 10 dB of boost or a 20 dB cut to the bass, mid-range, and treble. I recommend leaving the AF MIDDLE TONE GAIN at zero and applying a small amount of boost or attenuation to the AF BASS GAIN and/or AF TREBLE GAIN as you see fit. Boosting or cutting all three should be avoided as it will cause a lumpy audio passband.

Select <FUNC> <RADIO SETTING> *<select the mode>* and turn the FUNC knob or use the < > keys to adjust the value of each of the three, tone controls.

The default for all settings is zero, creating a flat audio passband.

PARAMETRIC EQUALIZER (TRANSMIT)

You can tailor the audio frequency response of your transmitted signal using the three-band parametric equalizer. There are settings for when you have the speech compressor turned off and for when you have the compressor turned on.

- Before setting up the microphone equalizer, you have to turn it on using <FUNC> <MIC EQ>.
- Set the SSB levels and the transmitter bandwidth, as per the previous section.
- First, we will set the equalizer with the speech compressor turned off. If necessary, press <FUNC> <PROC LEVEL> and use the FUNC knob to turn the level down to OFF. There should be no PROC indicator on the main display. Later we can set the other parametric equalizer with the speech compressor turned on.

- The parametric equalizer settings are at <FUNC> <OPERATION SETTING> <TX AUDIO>.

It is always preferable to make transmitter adjustments with the radio connected to a suitable high power 50 ohm load. A high power (100W) 50 dB or 60 dB, 50 ohm attenuator will work as well. The test setup is the same as for the SSB adjustments above. If you do not have a suitable load, you can set up the SSB levels while transmitting into your normal antenna. Preferably carry out the adjustment during the daytime when the HF bands are closed and announce your callsign regularly.

It is a good idea to listen to your transmitter audio quality using headphones. You can use the speaker in the radio, but headphones are much better for this task. Turn on the transmit monitor using <FUNC> < MONI LEVEL>. Turn up the monitor level with the FUNC knob. The audio for the transmit monitor is taken from the transmitter I.F. so it is a good representation of the signal that is being transmitted.

The equalizer settings are a personal preference. However, I would avoid using very narrow bandwidth settings because they will make the audio passband lumpy. You are aiming for a fairly linear slope across the audio passband. There is no point in placing the low-frequency filter below the low edge of the transmitter bandwidth or the high frequency above the high end of the transmitter bandwidth. The equalizer will have no effect on frequencies that are not within the band of frequencies being transmitted.

FREQ sets the center frequency of the low, mid, or high-frequency filter.

- LOW is adjustable from 100 Hz to 750 Hz
- MEDIUM is adjustable from 750 Hz to 1500 Hz
- HIGH is adjustable from 1500 Hz to 3200 Hz

LEVEL sets the gain or loss at the center frequency in dB (+10 to -20 dB)

BWTH sets the bandwidth of the filter (0 to 10)

➢ **Speech processor (compressor) turned off**

PARAMETRIC EQUALIZER SETTINGS – Compressor Off			
Control	Function	ZL3DW Setting	My Setting
PRMTRC EQ1 FREQ	LOW FREQ	300 Hz	
PRMTRC EQ1 LEVEL	GAIN	-3 dB	
PRMTRC EQ1 BWTH	Q (Bandwidth)	Q 5	
PRMTRC EQ2 FREQ	MID FREQ	900 Hz	
PRMTRC EQ2 LEVEL	GAIN	0 dB	
PRMTRC EQ2 BWTH	Q (Bandwidth)	Q 5	
PRMTRC EQ3 FREQ	HIGH FREQ	2400 Hz	
PRMTRC EQ3 LEVEL	GAIN	+8 dB	
PRMTRC EQ3 BWTH	Q (Bandwidth)	Q 5	

➢ **Speech processor (compressor) turned on**

The speech processor (compressor) only works in SSB mode, so the 'P PRMTRC' settings only work when you are using SSB, and you have the speech processor turned on. Press <FUNC> <PROC LEVEL> and use the FUNC knob to turn the level up to your normal compression level. PROC is indicated on the main display, at the top of the spectrum scope, below the ATT Soft Key.

PARAMETRIC EQUALIZER SETTINGS – Compressor On			
Control	Function	ZL3DW Setting	My Setting
P PRMTRC EQ1 FREQ	LOW FREQ	300 Hz	
P PRMTRC EQ1 LEVEL	GAIN	-6 dB	
P PRMTRC EQ1 BWTH	Q (Bandwidth)	Q 5	
P PRMTRC EQ2 FREQ	MID FREQ	900 Hz	
P PRMTRC EQ2 LEVEL	GAIN	0 dB	
P PRMTRC EQ2 BWTH	Q (Bandwidth)	Q 5	
P PRMTRC EQ3 FREQ	HIGH FREQ	2400 Hz	
P PRMTRC EQ3 LEVEL	GAIN	+8 dB	
P PRMTRC EQ3 BWTH	Q (Bandwidth)	Q 5	

SETTING UP THE VOICE MEMORY KEYER

The voice memory keyer is a very useful feature of the radio. It can be used for sending CQ calls or any information that you transmit often. It is especially useful for contest operation. Especially if the contest has a fixed 'contest exchange' such as the signal report and your CQ zone number. You can record the contest exchange, *"thanks you are 59 32"* as well as your *"CQ Contest."* Having one message with just your callsign in phonetics, *"Zulu lima three delta whisky"* is useful for when you are trying to break a pile-up, to work a rare DX station or a DXpedition.

There are five voice memory slots. Each recording can be up to 20 seconds long. It is longer than you think. My messages are all less than 10 seconds long.

Recording or sending the voice messages is done using the menu command <FUNC> <MESSAGE>. This brings up the 'Message Memory' popup. Unfortunately, the box disappears if you touch the screen or press any button and it overlays the center of the spectrum and waterfall display. I can see why investing in the FH-2 external keypad is a good idea.

TIP: Pressing the FUNC button when the message keyer popup is open sends the last used message. This means that you have to close the message keyer popup window before making any other selections with the FUNC control.

FH-2 keypad: You can use the FH-2 keypad, or similar, to send voice messages or set up a recording. You cannot use a USB keyboard.

Recording: The radio will not transmit while you are recording a voice message.

Touch the MEM Soft Key. A red REC indicator will flash under the mode indicator just to the right of the meter display. The option will time out after five seconds.

Touch one of the message keys, 1 to 5. When you are ready to record, click and hold the microphone PTT switch. The REC indicator will stop flashing and stay lit. Record your message in your 'normal' on-air voice. Release the microphone PTT to complete the recording. Do the same thing to record the other 4 macros. Touch BACK, or the screen outside of the popup window, or press any button to exit.

RX Level: Sets the volume of the message as it plays back. You will hear the recording even if you have MONI turned off. If it annoys you, turn the RX Level on the Message Memory screen down to zero. If BK-IN is turned off, you can playback the messages without transmitting.

TX Level: Sets the level to be transmitted. You should set it so that it is approximately the same as when you say the same message using the microphone. If I leave the TX Level at the default level of 50%, the COMP meter hardly flickers, but the ALC meter looks the same as when transmitting from the microphone.

If I set the TX level to 85% the ALC and COMP meters both look the same as when transmitting from the microphone. I left the TX Level at the default level of 50%.

Playback: Once the messages have been recorded, you can touch 1 – 5 to play them back. You can listen to your recordings by keying any of the messages with BK-IN turned off. If the RF/SQL knob has been set to act as a squelch control, the green BUSY light will come on while the message is playing.

You must have <FUNC> <BK-IN> turned on to transmit the messages. The red TX light will come on while the message is playing.

SETTING UP THE CW MESSAGE KEYER

There are two ways you can record messages for the CW keyer. You can type in the macro as TEXT from the on-screen or external USB keyboard, or you can record a MESSAGE sent with a CW Paddle.

Before you record or type the message you have to set the five message memory slots to TEXT or MESSAGE. You can have some slots set to TEXT and others set to MESSAGE. Select TEXT or MESSAGE for each of the five memory slots using <FUNC> <CW SETTING> <KEYER> <CW MEMORY>.

TIP: It could be handy to have one slot set to MESSAGE so you can use the paddle to quickly save another station's callsign, during or immediately before beginning a QSO.

TIP: If you have the FH-2 or compatible keypad, you can initiate a recording by pressing the MEM key. This is useful if you are using the MESSAGE mode and recording a message from the paddle.

➢ **Saving a CW keyer memory (TEXT method)**

To save **text** to a CW keyer memory slot. Make sure the message slot you want to use is set to TEXT. Press <FUNC> <CW SETTING> <KEYER> <CW MEMORY (1-5)> <TEXT>. You can have some slots set to TEXT and others set to MESSAGE.

- Set the radio to CW mode.
- Turn <FUNC> <BK-IN> off.
- Press <FUNC> <MESSAGE> <MEM> and select message 1-5.

A keyboard will appear on the display. Type in your message. You are allowed up to 50 characters. The last character must be the } symbol. It is used to tell the transceiver to finish transmitting and return to receiving. The arrow keys move you across the message. The 'X in an arrow' symbol ⊠ is the backspace. Touch ENT to save the message, or BACK to exit without saving.

TIP: To listen to the message without transmitting, set BK-IN to off and then touch the message number. BK-IN must be turned on for the message to be transmitted.

To insert a **contest number**, enter a # character. For example, 'R 5NN # K}'. The contest number will be sent as a three-digit number with leading zeros. The contest number will extend to four digits once 1000 is reached.

You can **decrement** the contest number by touching the DEC Soft Key. To set the number back to 1, or any other number, use <FUNC> <CW SETTING> <KEYER> <CONTEST NUMBER>.

The number format is set using the <FUNC> <CW SETTING> <NUMBER STYLE> command. See page 137 for the different number format options.

Morse code prosigns

For KN use	(	Over to a specified station
For AS use	&	Stand By
For AR use	+	End of message
For BT use	=	New section or paragraph
For CT use	%	New message
For SN use	!	Message understood

As far as I can tell, you cannot enter prosigns such as AA, BK, BN, CL, HH, or SK into a CW text memory. It should be possible to enter them from the paddle if you are using the MESSAGE method.

➢ **Saving a CW keyer memory (MESSAGE method)**

First, make sure the message slot you want to use is set to MESSAGE. Press <FUNC> <CW SETTING> <KEYER> <CW MEMORY (1-5)> <MESSAGE>.

To save a message sent by a CW paddle to a keyer memory slot. Set the radio to CW mode. Turn BK-IN off. Turn MONI on, unless you are using an external keyer with its own sidetone. Press <FUNC> <MESSAGE> <MEM>. Slots set to MESSAGE show a series of vertical bars instead of text. Touch to select one of them.

A red REC icon will flash. Start sending your message within 10 seconds and touch MEM to stop the recording.

TIP: To listen to the message without transmitting, set BK-IN to off and then touch the message number. BK-IN must be turned on for the message to be transmitted.

TIP: You could allocate one memory slot for editing during or immediately before a QSO to put in the other station's callsign.

➢ **Sending a message**

Once recorded, you can send the message by pressing 1-5 on the FH-2 keypad, or by opening the message memory popup, <FUNC> <MESSAGE> and touching 1-5. Like all CW operation, you must have BK-IN turned on. If the CW decoder screen is turned on, the text string is displayed on the bottom line of the decoder screen. Each character turns yellow as the text message is sent.

Pressing FUNC when the message popup is onscreen resends the last used message.

➢ **Stopping a message from being sent**

To stop a message while it is being sent, start sending from the paddle, touch the message memory key on the touchscreen, or press the message key on the FH-2 keyer.

GETTING READY FOR CW OPERATION

The main controls for CW operation are the keying speed and pitch. They can be adjusted at any time, to suit your needs. Set the R.FIL Soft Key for the 500 Hz roofing filter and use the FUNC menu to turn on BK-IN, MONI, and KEYER. The icons will be displayed on the top line of the spectrum display. I often choose a 50 kHz span for the spectrum scope as it provides a good balance between the range of displayed frequencies and the narrow receiver bandwidth.

➢ **Morse keys**

There are provisions for different types of Key. The rear panel 'Key' jack can be used with a Paddle, Bug, or Straight Key. Note that you must turn <FUNC> <BK-IN> on for the keyer to automatically place the transceiver into transmit mode. The more advanced CW menu settings are discussed in the CW SETTING menu on page 134.

➢ **CW Keyer**

You activate the internal keyer by setting the radio to CW mode and selecting <FUNC> <KEYER>. Once this has been done the keyer will be active anytime you select the CW mode.

See CW SETTING – KEYER sub-menu on page 137 for the full list of keyer settings.

➢ **Keying speed**

<FUNC> CW SPEED> sets the speed of the electronic keyer including CW sent from the CW message keyer. As you adjust the FUNC control, the CW speed in words per minute (wpm) is displayed on a popup in the middle of the display. The CW speed is adjustable from 4 wpm to 60 wpm.

> ➢ **CW pitch**

<FUNC> CW PITCH> changes the pitch (tone) of a received CW signal without changing the receiver frequency. Use the FUNC control to set the pitch so that CW sounds right to you. A popup displays the pitch frequency. It is adjustable from 300 Hz to 1050 Hz. Most people use a tone between 700 and 1000 Hz. I prefer 750 Hz.

> ➢ **Pitch offset on SSB**

CW signals are received at an offset equal to the 'pitch' setting of the CW PITCH control. The behavior when you switch from CW to SSB varies depending on whether you have selected the default 'pitch offset' or the 'direct frequency' option. Note that this setting only affects the VFO in SSB mode. In CW mode the VFO always shows the frequency of the received CW signal.

<FUNC> <CW SETTING> <MODE CW> <CW FREQ DISPLAY> (PITCH OFFSET or DIRECT FREQ)

The default PITCH OFFSET setting offsets the VFO when you switch from CW to SSB, so you hear the CW signal at the same tone.

If you select the DIRECT FREQUENCY setting, the VFO frequency will not be offset if you change from CW to SSB. The CW tone will be at the zero-beat frequency and will not be heard.

I believe that the default PITCH OFFSET setting is preferable. It means that you can tune in a CW signal while on SSB and then switch to CW without losing the signal.

> ➢ **Break-in setting (BK-IN Soft Key)**

Your CW Morse Code signal will not be transmitted unless <FUNC> <BK-IN> has been turned on. You can leave it on all the time. The setting is remembered when you change modes. With BK-IN turned off you will hear the CW signal, but it will not be transmitted unless the MOX button, the microphone PTT, or an external PTT signal perhaps from a footswitch, is activated.

There is no front panel control for selecting full or semi break-in. I guess most people use one or the other most of the time. Select SEMI or FULL Break-in using <FUNC> <CW SETTING> <MODE CW> <CW BK-IN TYPE>.

- Full break-in mode will key the transmitter while the CW is being sent and will return to receive as soon as the key is released. This allows for reception of a signal between CW characters. There is some relay clicking but it is not too intrusive.

- Semi break-in mode will key the transmitter while the CW is being sent and will return to receive after a delay when the key is released.

  The break-in delay is set using <FUNC> <BK-IN DELAY>.

> **Roofing filter**

Use the R.FIL Soft Key to set the roofing filter. It will normally be set to 500 Hz for CW.

> **Tuning bar graph**

A bar graph is displayed under the S meter display to indicate the tuning of the CW signal. When the signal is tuned perfectly, three white dots are displayed right in the middle, under the red dot.

You can turn the bar graph on or off with <FUNC> <CW SETTING> <MODE CW> <CW INDICATOR>.

> **Sidetone**

The sidetone level is adjusted by selecting <FUNC> <MONI LEVEL> and using the FUNC knob to set the transmit monitor level. Changing the monitor level on CW will not affect the monitor level when you are in SSB mode.

> **CW Keyer messages (keyer macros)**

There are five CW message memory slots which can be used for DX or Contest operation or just to save you sending the same message over and over. They are great for sending CQ on a quiet band. The CW keyer can send an auto-incrementing contest number. See 'setting up the CW message keyer' on page 19 in the previous section.

> **The filter function display**

The filter function display is a spectrum display that indicates the bandwidth of the currently selected roofing filter and other things such as the manual notch, contour control, APF filter position, I.F. width, and I.F. shift. The filter function display is hidden when the spectrum display is expanded, (EXPAND is blue).

> **APF filter**

Don't forget to use the fabulous APF (audio peak filter). It is an audio DSP filter centered on the CW pitch frequency. Because the APF is so narrow it is very effective at lifting weak CW signals out of the noise. In the CW mode, pressing the CONT/APF button once turns on the APF filter. Pressing it again turns on the Contour filter. And pressing it a third time turns the filters off. Turning the (outer) CONT/APF knob enables the APF filter and allows you to change the filter center frequency. If contour is selected, the same knob adjusts the center frequency of the contour filter.

The best way to ensure that you are using APF is to look at the filter function display. The APF is represented with a vertical orange line near the center frequency of the spectrum. The contour filter is represented with an orange dip or peak.

There are three APF bandwidths available. The wider bandwidth setting will allow for the VFO being slightly off tune, but in a contest situation, you could hear more than one CW signal. The narrow setting requires that the signal be tuned exactly, or you may "lose the signal." <FUNC> <OPERATION SETTING> <RX DSP> <APF WIDTH> (wide, medium, or narrow).

➢ **CW keying in SSB mode**

Some people like to use a CW sign-off when operating SSB. The CW AUTO MODE menu setting <FUNC> <CW SETTING> <MODE CW> <CW AUTO MODE> lets you enable the Morse key or paddle when you are on SSB. When set to ON the key is 'live.' When set to OFF, the key is disabled on SSB. If you select 50M the key is active on the 6m band but not on the HF bands. I suggest leaving this function turned off unless you really want the ability to send CW while in SSB mode. This function ignores the BK-IN setting. CW will be sent even if BK-IN is turned off.

➢ **QSK delay time**

<FUNC> <CW SETTING> <MODE CW> <QSK DELAY TIME> sets the time before the transceiver will start sending the CW signal. This allows your linear amplifier to switch before any power is transmitted.

The default is 15ms which should be fine for any modern solid-state amplifier. However, you can select a longer delay if your amplifier uses relay switching. If you are operating at more than 45 wpm the radio automatically selects 15 ms regardless of the menu setting.

RTTY AND PSK MESSAGE KEYERS

The RTTY and PSK message keyers only work in the internal PSK and RTTY modes. They are identical, and nearly the same as the CW keyer. There is no way to send a contest number and you use an 'End' ↵ character to finish the text string instead of the } symbol. A contest number would have been handy. I have no idea why the RTTY and PSK message keyers are different in this way.

There are five message memories. Each can hold 50 characters.

➢ **Saving or editing a message**

To save text to a PSK or RTTY keyer memory slot.

- Set the radio to PSK or RTTY-L mode.
- Press <FUNC> <MESSAGE> <MEM> and select message 1-5.

A keyboard will appear on the display. Type in your message. You are allowed up to 50 characters. The last character must be the ↵ symbol entered by touching 'End.' The 'End' ↵ symbol is used to tell the transceiver to finish transmitting and return to receiving. The arrow keys move you across the message.

The 'X in an arrow' ⊗ symbol is the backspace. Touch ENT to save the message, or BACK to exit without saving.

TIP: You could allocate one memory slot for editing during or immediately before a QSO to enter the other station's callsign.

➢ **Sending a message**

Once recorded, you can send the message by pressing 1-5 on the FH-2 keypad, or by opening the message memory popup, <FUNC> <MESSAGE> and touching 1-5.

If you are using the internal decoder, open that before you start the message popup. The text string is displayed on the bottom line of the decoder screen. Each character turns yellow as the text message is sent.

Unlike the voice keyer and the CW keyer, you do not have to have BK-IN turned on to transmit a saved message.

➢ **Stopping a message from being sent**

To stop the message while it is being sent, touch the same message key on the touchscreen, or press the message key on the FH-2 keyer.

GETTING READY FOR RTTY OPERATION

The radio supports three kinds of RTTY operation. Firstly, there is the onboard RTTY decoder. Which can be used with the RTTY message memories. The second method is to use external PC software such as MixW, MMTTY, DM780, or Fldigi with AFSK (audio frequency shift keying). AFSK uses two audio frequencies to create the frequency-shift keying in the SSB DATA-U mode. The third method is to use external PC software with FSK (frequency shift keying), which uses digital signals to key the transceiver to predefined mark and space offsets. FSK uses the radio's RTTY-L mode, so you can use the RTTY message memories.

➢ **Keyer send messages (keyer macros)**

There are five RTTY message menus which can be used for DX or Contest operation or even just to save you sending the same message over and over. They are great for sending CQ on a quiet band. The setup is on page 22 in the previous section. The keyer messages only work if you are in the RTTY mode, so they will work for the internal RTTY mode or if you are using an external RTTY program with FSK keying, but not if you are using the DATA-U mode and an external AFSK RTTY program.

➢ **Onboard RTTY operation**

If you can find any RTTY activity on the bands, using the onboard RTTY mode is simple. You have an onboard RTTY decoder and a set of five pre-defined messages.

Unfortunately, the operation of the transmit messages is not integrated with the decoder screen.

1. Press the MODE button or touch the mode icon and select the RTTY-L mode.

2. It is normal to set the roofing filter to 500 Hz, but you can choose the 3 kHz filter. Set the AGC to AUTO. This radio is an SDR so there is no need to turn off the AGC. The IPO Soft Key is usually set to AMP1. You could set it to IPO if you are using one of the three low bands, or AMP2 for the 10m or 6m band. ATT should be OFF.

3. Tune to an RTTY signal. The two audio peaks need to line up with the RTTY mark and space lines (M S) on the filter function display.

4. Turn on the decode screen. <FUNC> <DECODE>. This must be done before activating the RTTY keyer. The decoder only has two controls. The DEC OFF Soft Key closes the window. DEC LVL sets the decoding threshold. Touch the Soft Key and adjust the DECODE LVL with the FUNC knob to reduce the amount of rubbish text being displayed. You have to be very quick. The FUNC control reverts to whatever it was on previously after two seconds.

   The four blank Soft Keys could have been used for sending RTTY messages, but they are not.

5. Below the received text area there is one line reserved for displaying the outgoing text. The characters turn yellow as they are sent.

6. If you have the FH-2 external keypad (or a copy), you should use that to send the message macros. Otherwise, turn on the RTTY message keyer. <FUNC> <MESSAGE>. Touch 1 to 5 to send the appropriate message.

The onboard decoder and message keyer are not very well implemented. Firstly, the decode window covers the spectrum scope so that you cannot see RTTY signals on the band, and secondly, the message keyer is not integrated into the decoder. The message keyer window covers the received text so you can't see the message that you are responding to. Touching any part of the screen outside of the popup window or pressing some of the buttons closes the popup, and you have to go back through the FUNC menu to get it up again.

TIP: If the received signal is strong but the decoder is printing gibberish, the other station might be transmitting on the wrong sideband. Switch to RTTY-U to decode the signal. Amateur band RTTY should be on RTTY-L. Commercial Shortwave band RTTY is likely to be received on RTTY-U.

➢ **AFSK RTTY from an external PC program**

To use an external PC based digital modes program for transmitting you use the DATA-U mode, not the RTTY mode. Press the MODE button and select DATA-U.

The easiest way to send RTTY from your favorite external digital mode program is to use AFSK rather than FSK keying. With AFSK, the RTTY signal is sent to the transceiver as audio tones rather than as a digital signal. I have included setup instructions for several digital modes programs in the 'Setting up PC and digital mode software' section.

First of all, you have to have a connection between the transceiver and the PC. The easiest and most modern method is to use a single USB cable. It will carry the CAT control commands, com port lines for PTT, and audio. See the section about 'Setting up a USB connection,' on page 32. It only has to be done once. You can also use the old-fashioned separated RS-232 communications and audio cables method, but I have not covered the setup for that.

➤ AFSK RTTY audio levels

Once the digital mode software can communicate with the radio, it is time to set the transmitted and received audio levels. You can set the audio level for the transmitter in the FTDX10, or you can set the levels using the soundcard mixer controls in your PC, or possibly with level controls in the digital mode software.

<FUNC> <RADIO SETTING> <MODE PSK/DATA> <RPORT GAIN>

The levels should be set so that the transmitter power is a fraction under 100 watts, (or a lower power level). That way there is no risk of overdriving the transceiver and causing distortion. While transmitting, the ALC meter should be low or zero, indicating that the automatic leveling is not working very hard. High levels of ALC could compress the transmitted audio causing the transmission to be noisy and less easy to copy. I found the MixW output level to be quite high, so I set RPORT GAIN to 12 for RTTY and the Windows Sound Mixer to 10. You can also use the 'sound' settings inside the digital mode software to adjust the transmitted level.

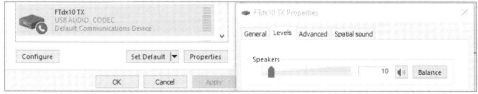

There is no control in the radio for setting the level of the receiver audio output to the PC via the USB cable. You have to use the Windows 'sound settings' software and/or the 'sound' settings inside the digital mode software.

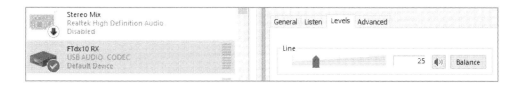

➢ **FSK RTTY**

I have included instructions for setting up MMTTY for FSK RTTY in the section titled, 'Setting up PC digital mode software,' (page 39). MMTTY is often used in conjunction with the N1MM logger. FSK also works with Fldigi. I cannot get FSK to work with DM780, the digital modes program that is used by Ham Radio Deluxe. However, CW and the AFSK RTTY mode both work OK.

➢ **RTTY Mark and Shift frequencies**

You can change the RTTY MARK tone frequency and the RTTY shift. The default Mark tone is at 2125 Hz and the Shift when using RTTY-L is +170 Hz.

You can change the settings, but I don't see any reason to do it.

<FUNC> <RADIO SETTING> <MODE RTTY> <MARK FREQUENCY>

<FUNC> <RADIO SETTING> <MODE RTTY> <SHIFT FREQUENCY>

➢ **RTTY polarity settings**

You can change the RTTY polarity settings for the receiver and for the transmitted RTTY signal. The default for amateur radio is to leave them set to NOR (normal).

<FUNC> <RADIO SETTING> <MODE RTTY> <POLARITY TX>

<FUNC> <RADIO SETTING> <MODE RTTY> <POLARITY RX>

With a NOR setting the VFO indicates the Mark frequency, and the data signal causes transitions to the Space frequency.

With a REV setting the VFO indicates the Space frequency, and the data signal causes transitions to the Mark frequency.

➢ **USB cable settings**

<FUNC> <RADIO SETTING> >MODE PSK/DATA>. Set REAR SELECT to USB for the audio input, and RPTT to RTS for the PTT control.

➢ **NOR, REV, RTTY-L, and RTTY-U**

It is easy to confuse the normal and reverse polarity settings set in the RADIO SETTING menu with the RTTY-L and RTTY-U settings selected with the MODE button.

For amateur radio, we use RTTY-L and the NOR settings. Commercial Shortwave band RTTY is likely to be RTTY-U.

If a station is transmitting RTTY on the wrong sideband you can switch to RTTY-U or REV to receive the signal correctly.

RTTY Settings		
Function	Operation	Tones
NOR	The VFO indicates the Mark frequency	FSK data shifts the frequency from the Mark tone to the Space tone
REV	The VFO indicates the Space frequency.	FSK data shifts the frequency from the Space tone to the Mark tone
RTTY-L	Mark 2125 Space 2295 (+170)	Transmits Mark low, Space high
RTTY-U	Mark 2295 Space 2125 (-170)	Transmits Mark high, Space low

GETTING READY FOR PSK OPERATION

The radio supports two kinds of PSK operation. Firstly, there is the onboard PSK decoder, which can be used with the PSK message memories. The second mode is to use external PC software such as MixW, MMTTY, DM780, or Fldigi.

The advanced menu settings for PSK operation are included in the chapter about the FUNC menu commands, so I won't repeat them here. See,

<FUNC> <RADIO SETTING> <MODE PSK/DATA> (on page 127), and

<FUNC> <RADIO SETTING> <ENCDEC PSK> (on page 131)

Select the PSK mode by pressing the MODE button. You may notice that there is no upper or lower sideband selection because BPSK (binary phase-shift keying) is sent using 180-degree phase changes rather than tones like RTTY and the transmitted sideband does not matter. However, QPSK (quadrature phase-shift keying) must be received on the same sideband it was transmitted on, so there are normal and reverse polarity settings in the menu structure. <FUNC> <RADIO SETTING> <ENCDEC PSK> (on page 131)

BPSK-31 and QPSK-31 are supported by the internal PSK encoder and decoder. BPSK-63 and QPSK-63 are not. This is not a big problem since PSK-63 is rarely used anyway.

If you want to change the PSK tone for the internal PSK mode, select <FUNC> <RADIO SETTING> <MODE PSK/DATA> <PSK TONE>. You can choose from 1000 Hz (default), 1500 Hz, or 2000 Hz. This setting only affects the audio frequency you hear when you tune into a PSK signal. It is like the audio offset applied to a CW signal. The PSK signal will be received and transmitted on the displayed VFO frequency regardless of this setting.

➢ **Onboard PSK operation**

Using the onboard PSK mode is simple. You have an onboard PSK decoder and a set of five pre-defined messages. Unfortunately, the operation of the transmit messages is not integrated with the decoder screen.

1. Press the MODE button or touch the mode icon and select the PSK mode.

2. It is normal to set the roofing filter to 3 kHz while tuning the band and then 500 Hz when you are in a QSO. Set the AGC to AUTO. This radio is an SDR so there is no need to turn off the AGC. The IPO Soft Key is usually set to AMP1. You could set it to IPO if you are using one of the three low bands, or AMP2 for the 10m or 6m band. ATT should be OFF.

3. Tune the PSK signal so that the PSK audio peak lines up with the C carrier point marker on the filter function display.

4. Turn on the decode screen. <FUNC> <DECODE>. This must be done before activating the PSK keyer. The decoder only has two controls. The DEC OFF Soft Key closes the window. DEC LVL sets the decoding threshold. Touch the Soft Key and adjust the DECODE LVL with the FUNC knob to reduce the amount of rubbish text being displayed. You have to be very quick. The FUNC control reverts to whatever it was on previously after two seconds.

   The four blank Soft Keys could have been used for sending PSK messages, but they are not.

5. If you have the FH-2 external keypad or a copy, then use that to send the message macros. Otherwise, turn on the PSK keyer popup. <FUNC> <MESSAGE>. Touch 1 to 5 to send the appropriate message.

The onboard decoder and message keyer are not very well implemented. Firstly, the decode window covers the spectrum scope, and secondly, the message keyer is not integrated into the decoder. The message keyer popup window covers the received text. Touching any part of the screen outside of the popup or pressing some of the buttons closes the popup and you have to go back through the FUNC menu to get it up again.

➢ **PSK message keyer (keyer macros)**

There are five PSK message menus which can be used for DX or Contest operation or even just to save you sending the same message over and over. I have already covered the setup back on page 22. The PSK keyer messages only work if the radio is in the PSK mode. The inability to send messages from a connected keyboard, or even enter the other station's callsign without saving it to one of the five memory slots is rather limiting. So, I expect most people will use an external digital modes program for general PSK operation.

➤ **PSK from an external PC program**

When you use an external PC based digital modes program for transmitting PSK you select the DATA-U mode, not the PSK mode. Press the MODE button or touch the mode icon and select DATA-U.

There must be a connection between the transceiver and the PC. The easiest and most modern method is to use a single USB cable. It will carry the CAT control commands, COM port lines for PTT and keying, and the audio. See the section about 'Setting up a USB connection,' on page 32. The setup only has to be done once.

➤ **PSK audio levels**

Once the digital mode software can communicate with the radio, it is time to set the transmit and receive audio levels. Depending on your digital mode software, the RPORT audio level for PSK may need to be higher than for RTTY or FT8.

You can set the audio level for the transmitter in the FTDX10, or you can set the levels using the soundcard mixer controls in your PC or possibly with level controls in the digital mode software.

<FUNC> <RADIO SETTING> <MODE PSK/DATA> <RPORT GAIN>

The levels should be set so that the transmitter power is a fraction under 100 watts, and occasional peaking to full power when text is being sent. That way there is no risk of overdriving the transceiver and causing distortion. While transmitting, the ALC meter should be low or zero. I set RPORT GAIN to 12, and the Windows Sound Mixer to 10. You can also use the 'sound' settings inside the digital mode software to adjust the transmitted level.

There is no control in the radio for setting the level of the receiver audio output to the PC via the USB cable. You have to use the Windows 'sound settings' software or the 'sound' settings inside the digital mode software.

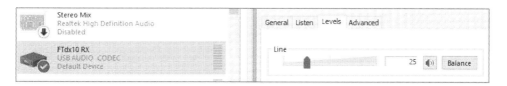

➤ **USB cable settings**

<FUNC> <RADIO SETTING> >MODE PSK/DATA>. I set REAR SELECT to USB, RPORT GAIN to 12, and RPTT to RTS.

I included typical COM port settings for a selection of popular digital mode programs in the 'Setting up PC digital mode' section on page 39. I can't include them all, but the settings I describe should be representative of what you need.

➤ **PSK AFC setting**

The internal PSK decoder has AFC (automatic frequency control) to help you tune the PSK signal accurately. It will pull the receiver slightly to get the signal tuned in. <FUNC> <RADIO SETTING> <ENCDEC PSK> <DECODE AFC RANGE> sets the bandwidth for the AFC action. A BPSK-31 signal has a bandwidth of about 31 Hz. So, I would experiment with the default 15 Hz setting or the 30 Hz setting.

➤ **BPSK and QPSK compared**

You can choose from BPSK (binary phase-shift keying) which is standard PSK-31, or QPSK (quadrature phase-shift keying). I assume that it is QPSK-31 since there is no baud rate control offered. QPSK has the advantage of having built-in error correction, but it is considerably less sensitive than BPSK, and much less popular on the bands. <FUNC> <RADIO SETTING> <ENCDEC PSK> <PSK MODE>.

Binary PSK transmits information by changing the phase of the carrier signal. A digital one is represented by a 180-degree phase change and a digital zero is represented by no phase change. The 31 baud rate of PSK-31 is chosen to provide a narrow bandwidth transmission at a typical typing speed. These days there is little PSK activity since most of the world has moved to FT8. The faster PSK-63 mode was less popular because you need a higher received signal to noise ratio to get error-free reception. In other words, it is less sensitive. It also uses twice the bandwidth.

QPSK is much faster than BPSK because each of the four-phase states carries two data bits of information. The higher data rate allows time to insert some additional 'information bits' to enable the use of forward error correction, so you tend to get nearly perfect decoding or none at all. QPSK requires more signal level for good decoding. It has never really caught on as a general chat mode, and most people that do use it prearrange their contacts. Typically, QSPK is transmitted on frequencies just above the PSK segment of the band.

BPSK-31 is a narrow bandwidth mode like CW, so it works well at low signals strengths. Not as well as FT8 though. It also works well on radio paths with fading or trans-polar paths with auroral flutter and multi-path effects. Ideal for 'long path' contacts between New Zealand and northern Europe.

GETTING READY FOR FT8 AND OTHER DIGITAL MODES

The setup for FT8 or other digital modes is the same as setting the radio up for 'PSK from an external PC program,' or 'AFSK RTTY from an external PC program,' covered above. The method for setting the audio levels is the same as well.

You need to set up a USB connection between the radio and your PC as described in the next section 'Setting up a USB connection.' Then you need to set up the software on the PC. There are hundreds of software options. I have included basic setup instructions for some of the most popular packages in the 'Setting up PC digital mode software,' starting on page 39.

PRESET

The April 2021 firmware release added a 'Preset' mode to the MODE selection. The idea is to provide an easy way of changing the transceiver settings for FT8 or any other digital mode, then changing it back when you have finished. When the PRESET Soft Key is blue the preset settings are being applied. When the PRESET Soft Key is grey the preset settings are not being applied.

TIP: The FTDX10 only has one Preset. You could configure it with your FT8 settings and configure the non-preset 'MODE PSK/DATA' settings for a different mode.

You should still use the DATA-U mode. The preset does not include mode switching. Press the MODE button or touch the mode icon to set the mode, to DATA-U for FT8. Then press <MODE> <PRESET> to turn on the PRESET. Touch and hold the PRESET Soft Key to change the 14 preset settings.

Function	PRESET Settings (default)	ZL3DW	My settings
CAT RTS	**OFF** or ON	OFF	
CAT RATE	4800 / 9600 / 19200 / **34800**	38400	
CAT TIME OUT	**10ms**/100ms/1000ms/3000ms	10 ms	
LCUT FREQ	OFF / **100 Hz** to 1 kHz	100 Hz	
HCUT FREQ	OFF 700 to 4000 Hz (**3200 Hz**)	3200 Hz	
TX BPF SELECT	**50-3030 Hz** / 100-2900 Hz/ 200-2800 Hz/ 300-2750 Hz/ 400-2800 Hz	100-2900 Hz	
REAR SELECT	DATA/**USB**	USB	
RPORT GAIN	0 to 100 (**10**)	12	
RPORT SELECT	DAKY / **RTS** / DTR	RTS	
AGC FAST DELAY	20 – 4000 ms (**160 ms**)	160 ms	
AGC MID DELAY	20 – 4000 ms (**500 ms**)	500 ms	
AGC SLOW DELAY	20 – 4000 ms (**1500 ms**)	1500 ms	
LCUT SLOPE	6 dB/octave or **18 dB/octave**	18 dB	
HCUT SLOPE	6 dB/octave or **18 dB/octave**	18 dB	

The preset setup uses a standard menu format, which makes me wonder why Yaesu didn't add a PRESET setup menu to the RADIO SETTING tab on the FUNC menu, instead of making it the only touch and hold Soft Key on the MODE selection screen.

SETTING UP A USB CONNECTION

A USB connection is the easiest way to establish audio pathways, CW and PTT signaling, and CAT (Computer Aided Transceiver) control of the radio from a computer. One 'USB 2.0 Type-A to Type-B' cable replaces all the old-fashioned interface boxes, audio cables, and RS232 connections that you used to use. Connect the USB cable between the USB Type-B port on the rear of the radio and any USB port on the computer. Feel free to use a spare USB 2.0 port on your computer as there is no advantage in wasting a USB 3.0 port.

A USB Type-B to USB Type-C cable should work, but I haven't tried it. The manual only refers to a 'Commercially available USB (A-B)' cable.

Using the USB cable for the first time is **not** "plug-n-play." You must download and install the Yaesu virtual COM port driver software. If you have already installed driver software for another recent Yaesu transceiver, you will not have to load it again.

➢ **Driver software**

The Yaesu driver software can be downloaded from the Yaesu website https://www.yaesu.com/.

Click on the FTDX10 image, or select the FTDX10 in the 'Products,' 'HF Transceivers/Amplifiers' tab and click the FTDX10 image. Click on the 'Files' sub-tab. In the 'Amateur Radio \ Software' tab, click and download the file labeled 'FTDX10 USB Driver (Virtual COM Port Driver).' The link will download 'CP210x_Windows_Drivers.zip.'

In Windows, click 'Extract All' to unzip the files and save them to a suitable place on your computer. I created a 'Yaesu FTDX10' folder under 'My Documents.'

Once the files have been unzipped, you can delete the ZIP file. It is only taking up hard drive space.

Navigate to the location where you saved the driver files. For Windows 10, double click the file called CP210xVCPInstaller_x64.exe to start the installation. It will install the 64-bit version of the driver. Follow the onscreen instructions to complete the installation. If you are running a 32-bit x86 operating system run the 32-bit installer instead.

After the driver has been installed. Connect the PC to the transceiver using the USB cable and turn the radio on. You should get the usual device connected 'bong' from Windows.

In Windows 10 select 'Settings' and then 'Devices.'

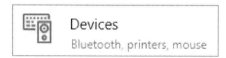

Under the 'Other Devices' heading, you should see a new device with the rather catchy name of, 'CP2105 Dual USB to UART Bridge Controller.' If the device is not listed, the PC is not seeing the radio. Try unplugging the USB cable from the PC end and then plugging it back in again. If it still won't work, reload the Yaesu driver software, with the USB cable connected between the radio and the PC and the radio turned off.

The driver software creates two COM ports. You will need to know the COM port numbers when you set up digital mode or other PC software. The 'Enhanced' COM port is used for CAT control of the radio. The 'Standard' COM port can't be used for CAT control, but its RTS and DTR lines are used to make the radio transmit (PTT) and to send CW or FSK RTTY data.

Start Windows 'Device Manager' by typing 'Device Manager' into the 'Find a setting box' on the same Windows screen, or into the Windows search box next to the 'Start' icon. Then click 'Device Manager.' Expand the 'Ports (COM & LPT)' line, to show the installed COM ports.

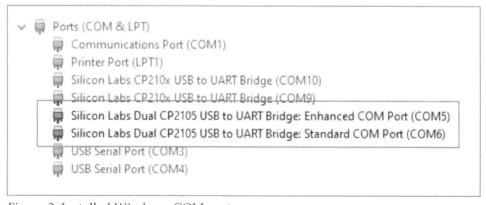

Figure 3: Installed Windows COM ports

The two COM ports of interest are the 'Silicon Labs Dual CP2105 USB… **Enhanced** COM Port' and the 'Silicon Labs Dual CP2105 USB… **Standard** COM Port.' Make a note of the COM port numbers and make sure you know which one is the Enhanced port.

> **Audio Codec**

In addition to creating the two USB ports, the driver software creates a USB audio CODEC (coder-decoder). This makes the radio look like an audio device, like a microphone or speakers.

In Windows 10 select 'Settings' and then 'Devices.'

Audio

⊲)) Speakers (Realtek High Definition Audio)

⊲)) USB Audio CODEC

You should see the Yaesu audio codec listed as 'USB Audio CODEC.' If it is not there, the sound won't work.

Click on 'Sound Settings' (right side of the window), then.

Click on 'Sound Control Panel' (right side of the next window).

Transmitter audio: The **Playback** tab controls the audio output from the PC into the transmitter. You should see a device labeled as 'USB Audio CODEC Speakers' or something similar. Double click the icon to open its properties.

I found the labeling confusing when trying to pick the right device in my digital

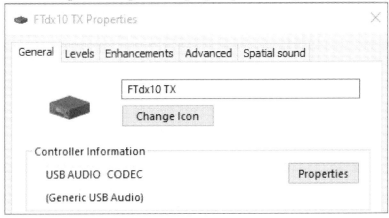

Figure 4: Renaming the Playback tab

modes programs, so I changed the name to 'FTdx10 TX,' and I changed the icon to one that looks like a transceiver.

Click on the 'Levels' tab and reduce the audio level to about 10. The transceiver input is very sensitive. Make sure there are no 'Enhancements' or 'Spatial Sound' effects selected. On the 'Advanced' tab, select 16 bit 48000 Hz (DVD Quality) and check both of the 'Exclusive mode' check boxes. Press the 'OK' button to return to the 'Sound' window.

Receiver audio: The **Recording tab** controls the audio output from the receiver to the computer. You will see a device labeled as 'USB Audio CODEC Microphone' or something similar. I found the labeling confusing when trying to pick the right device in my digital modes programs, so I changed the name to 'FTdx10 RX.'

Double click the icon and change the name on the General tab. You can select 'Change Icon' to pick a different icon if you like. I picked an icon that looks a bit like a transceiver.

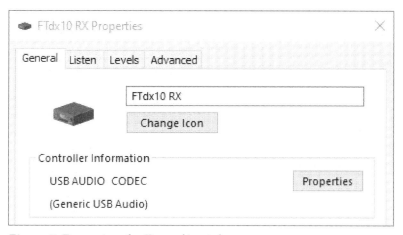

Figure 5: Renaming the Recording tab

Click on the 'Levels' tab and set the audio level to about 30. There is no control in the radio for setting the audio level to the computer, so you might have to change it later. On the 'Listen' tab make sure that 'Listen to this device' is <u>not</u> checked. The dropdown should be on 'Default Playback Device.'

The default setting on the Advanced tab should be 2 channel 16 bit 48000 Hz (DVD Quality) or 2 channel 44100 Hz (CD Quality). If you have a problem, it could be set to a '1 channel' option. The receiver output is on the Left audio channel only.

TIP: the transceiver volume control does not affect the audio level sent to the computer over the USB cable, but the squelch control is active.

➢ **USB cable settings for the CW mode**

If you want to key the transceiver in CW mode from external software.

Set <FUNC> <CW SETTING> <MODE CW> <PC KEYING> to DTR

➢ **USB cable settings for the DATA and PSK modes**

The following settings relate to the USB cable connection when using the DATA modes. They only need to be set once. See page 127 for a detailed explanation.

<FUNC> <RADIO SETTING> <MODE PSK/DATA> <DATA MOD SOURCE> <REAR>

<FUNC> <RADIO SETTING> <MODE PSK/DATA> <REAR SELECT> <USB>

<FUNC> <RADIO SETTING> <MODE PSK/DATA> <RPORT GAIN> <12>

<FUNC> <RADIO SETTING> <MODE PSK/DATA> <RPTT SELECT> <RTS>

Note: these settings will not be used if PRESET is turned on. See page 31.

➢ **USB cable settings for the PSK mode**

The settings for PSK are the same as the DATA mode settings above. If you have set them, you are already done. See page 127 for a detailed explanation.

➢ **USB cable settings for the RTTY mode**

There are no audio settings for FSK RTTY because it is a keyed mode (like CW).

<FUNC> <RADIO SETTING> <MODE RTTY> <RPTT SELECT> <RTS> sets the PTT keying line to RTS.

<FUNC> <RADIO SETTING> <MODE RTTY> <RTTY OUT LELEL> can be ignored. It sets the RTTY audio output level to the RTTY/DATA jack on the rear panel.

The mark and space frequencies and the RX and TX polarity settings can be left at their default settings. See page 26 for more details.

➢ **Radio settings for the SSB mode (external voice)**

The following settings relate to the USB cable connection when using the SSB modes. They only need to be set once. See page 119 for a detailed explanation.

SSB should not be used for digital modes from an external program, because the parametric microphone equalizer, VOX, and speech compressor might be active. Use the DATA-U mode instead.

<FUNC> <RADIO SETTING> <MODE SSB> <SSB MOD SOURCE> <REAR>

This setting would only be used if you were sending speech from the PC. Otherwise, leave it set to MIC. Note that the microphone will work normally even if this is set to REAR. But the voice keyer will not work. With BK-IN on, the radio will go to transmit, but the message will not modulate the transmitter.

<FUNC> <RADIO SETTING> <MODE SSB> <REAR SELECT> <USB>

<FUNC> <RADIO SETTING> <MODE SSB> <RPORT GAIN> <50> (set as required)

<FUNC> <RADIO SETTING> <MODE SSB> <RPTT SELECT> <RTS>

➢ **Other CAT settings**

Three other menu settings also affect CAT operation.

CAT RATE is used to set the interface speed of the USB connection between the radio and the PC software.

The speed should match the speed specified in the software application. It does not need to be very fast. 9600 bps (bits per second or 'bauds') is adequate for most connections. The default rate is 38400 bps. Your external digital mode software or any software that is communicating with the transceiver should be set for the same baud rate.

<FUNC> <OPERATION SETTING> <GENERAL> <CAT RATE> (38400 bps)

CAT TIME OUT TIMER sets the time that the radio will wait for a response to a CAT command. Leave it at the default 10 ms setting unless you are experiencing CAT control problems.

<FUNC> <OPERATION SETTING> <GENERAL> <CAT TIME OUT TIMER> <10>

CAT RTS is not the PTT control line that is set in the Radio Settings menu using RPPT SELECT. In this case, the COM port RTS line is used as a 'com interrupt.' When it is set to 'ON' the radio monitors the RTS line on the Enhanced COM port (CAT port) and will respond to changes initiated by the PC software. When it is set to 'OFF,' the radio does not monitor the RTS line on the Enhanced COM port.

<FUNC> <OPERATION SETTING> <GENERAL> <CAT RTS> <OFF> (default ON)

TIP: Some PC software does not like having CAT RTS turned on. Neither WSJT-X nor MixW will control the radio if CAT RTS is turned on. The Preset turns it off by default.

➢ **COM Port settings in PC software**

The COM port settings for the PC are usually set in each digital mode or control program. They can also be set in Device Manager, but that does not seem to be necessary. These are the settings I use.

Enhanced Port (used for CAT control)

- Port   COM x (the Enhanced CAT com port)
- Baud Rate   38400
- Data bits   8
- Parity   None
- Stop bits   1
- RTS   OFF, High, Hardware, Handshake, or PTT

- DTR   OFF, High, Hardware, Handshake, or CW

RTS and DTR must <u>not</u> be set to ON or always low.

Standard Port (used for PTT and CW signaling)

- Port   COM x (the Standard CAT com port)
- Baud Rate   9600
- Data bits   8
- Parity   None
- Stop bits   1
- RTS   PTT (must be the same as the RPTT setting)
- DTR   CW (must be the same as the PC KEYING setting)

RTS and DTR must <u>not</u> be set to ON or always low.

The RTS and DTR labels don't matter. You can use either line for the transmit PTT as long as you use the other line for CW. The names relate to old-fashioned RS-232 communications between 'old school' computers. RTS stands for 'ready to send' and DTR stands for 'data terminal ready.' But the lines have not been used for that sort of signaling since the 1970s.

*Anyway, I use 'ready to send' for the 'PTT' command because when PTT is active, you are ready to send the digital or CW signal. That leaves 'data terminal ready' for the CW or RTTY **data** signal.*

➢ RPORT transmitter audio settings

The RPORT GAIN settings control the audio level into the transmitter from the USB port or the RTTY/Data jack. You should turn it down so that the transmitter is just reaching full power (or less) when you are transmitting the digital mode text. ALC meter reading should be low or zero. For speech, it may be higher, but within the white zone.

There is an RPORT GAIN control for PSK/DATA (page 127), SSB (page 119), AM (page 121), and FM (page 124). There is no RPORT GAIN control for CW or the internal RTTY mode because they are 'keyed' modes.

TIP: as already discussed, the Windows sound 'playback' level should be adjusted before adjusting the RPORT GAIN. On my PC it is set to 10. Your PC may be different. There is a very good reason to turn down the Windows sound playback audio rather than just using the RPORT gain control.

If the level in the computer is too high the digital mode signal can create harmonics of the audio signal, which can result in the transmission of your digital mode signal on two frequencies at the same time. This understandably causes confusion and might make you unpopular. This is the reason that WSJT-X automatically transmits FT-8 on a 'Split' frequency. Shifting the transmitter frequency ensures that the modulating audio will be between 1500 to 2000 Hz and any audio harmonics cannot pass through the transmitter sideband filter.

➢ **USB cable receiver audio settings**

There are no controls for setting the receiver output level to the computer. So, you have to use the controls in the PC software and/or the Windows sound 'Recording' tab settings.

The receiver volume controls do not affect the audio level being sent over the USB cable to the computer software. But the squelch does. This means that you cannot have the receiver squelched and still send audio to the computer. I find this a bit disappointing. I would prefer that the audio to the computer over the USB cable be independent of the squelch but there does not seem to be any way to change it.

SETTING UP PC DIGITAL MODE SOFTWARE

Having set up a USB connection between the computer and the radio you will need to configure the digital modes or other software. This can be tricky and naturally, it is not covered in the Yaesu manual. In this section, I have included setup information that "works for me" for some popular digital mode and control programs. But of course, it is impossible to cover them all, and I have not covered all the settings, just the basic COM port ones. If you get stuck. Try Googling the application name and FTDX10 and hopefully some help will be on the way.

There are also some very good YouTube clips and the Yaesu users groups at https://groups.io/groups. Some of my digital modes software does not have a setting for the FTDX10 and this can be a big problem because the CAT commands are not always the same. I have had success using the FT-5000 or FTDX101 settings.

➢ **Downloads**

I strongly advise you to download the latest version of any software you want to use with the FTDX10. It can save you hours of frustration especially if the software has recently been updated to work with the transceiver. If it still doesn't work, at least you know you are using the latest version when you go online to seek help.

➢ N1MM Logger+

I am using N1MM Logger+ V 1.0.9221. It has support for the FTDX10. To set up communications with the radio, start N1MM with the transceiver turned on.

Enhanced Com port: Under 'Config - Configure Ports, Mode Control, Winkey, etc.'

- Set the top Port dropdown list to the Yaesu Enhanced COM port.

- Set the Radio dropdown to FTDX10.

- Uncheck CW/Other because you are not using this port for PTT.

- Click 'Set.'

- I have speed = 38400 (same as the radio), Parity = N, Bits = 8, Stop = 1, DTR = Handshake, RTS = Handshake. Radio = 1. Nothing else is selected.

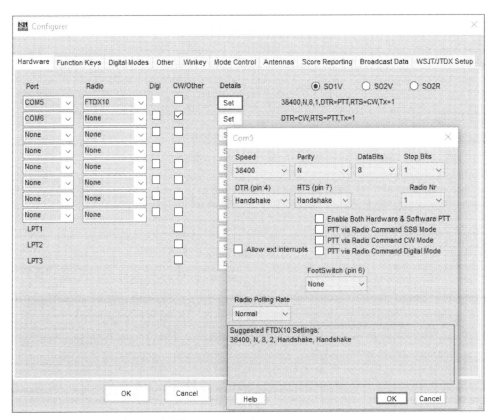

Figure 6: N1MM Enhanced COM port setting

Standard Com port: Under 'Config - Configure Ports, Mode Control, Winkey, etc.'

- Set the second Port dropdown list to the Yaesu Standard COM port.

- Set the 'Radio' dropdown to None.

- Do not check 'Digi,'
- Check 'CW/Other' because you will be using this port for PTT and CW.
- Click 'Set.'

- I have set DTR = CW, RTS = PTT, VFO = 1, PTT Delay = 30 ms. Nothing else is selected.

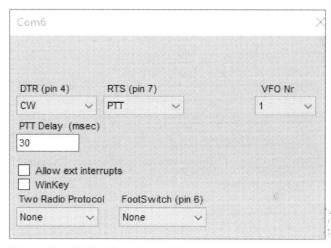

Figure 7: N1MM Standard COM port setting

➢ **CW Skimmer**

CW Skimmer uses OmniRig for CAT control.

➢ **OmniRig**

The latest V1.19 version of OmniRig does not have support for the FTDX10, but it does have support for the FTDX101. You have to download and install the software and also download the INI files.

http://www.dxatlas.com/Download.asp

Copy the FTDX101D.ini file into the 'Rigs' subfolder folder under the Afreet\OmniRig folder. You may be able to use the RIG 2 settings for PTT and CW control, but I have not tried it.

I have Radio = FTDX101D, Enhanced COM port, Baud Rate = 38400 (same as the radio), Data Bits = 8.

Parity = None, Stop = 1, RTS = Handshake, DTR = High, poll interval = 500 ms, Timeout = 4000 ms.

➢ **WSJT-X (FT8) (FT4) (JT65)**

There is an excellent WSJT-X user guide that tells you all about operating FT8 and the other WXJT modes, at http://physics.princeton.edu/pulsar/K1JT/wsjtx-doc/wsjtx-main-1.9.1.html#_standard_exchange.

Here are the settings that will get you started with WSJT-X. I am using V2.4.0.

- On the WSJT-X software, go to 'File' 'Settings' 'General' and enter your callsign, six-digit Maidenhead grid, and IARU region.

- Switch to the 'Radio' tab and set Rig to Yaesu FTDX10. (Only V1.9 and later versions of WSJT-X have the FTDX10 option).

  o Under the 'CAT Control' heading, set the Serial Port to the Enhanced Com port number. The **Baud Rate** should be the same as set in the radio. Default is 38400.

  o Data bits: **Default** or Eight, Stop bits: **Default** or One, Handshake: **None**.

  o Leave both 'Force Control Lines' dropdown boxes blank.

  o On the right 'PTT Method' side, set the Port to the radio's Standard COM port number.

  o Set the PTT method to RTS, Mode to Data/Pkt, and Split to 'Rig.'

  o Click 'Test CAT,' the button should turn green. If it does not. Check the baud rate on the radio and WSJT-X.

  o Click 'Test PTT,' the button should turn red and the transmitter should go to transmit, but no transmitter power should be created. Don't forget to click the button again to turn off the transmitter.

  o Click OK to exit.

- At this stage, the radio frequency should be indicated on the WSJT display and it should change if you turn the radio's VFO knob. You should be able to click the band dropdown beside the frequency display to select a band, and the radio should follow your selection.

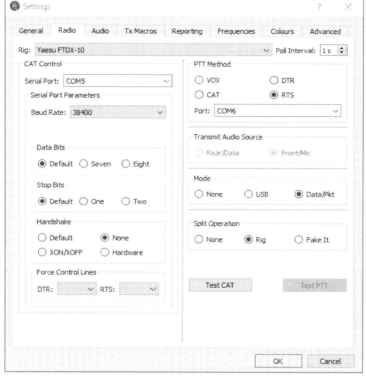

Figure 8: WSJT COM port settings

Go to 'File' 'Settings' 'Audio'

- Set the 'Input' to your FTDX10 RX audio output. Set the other setting to 'Left.'

- Set the 'Output' to your FTDX10 TX audio input. Leave the other setting at Mono or Left. It does not matter.

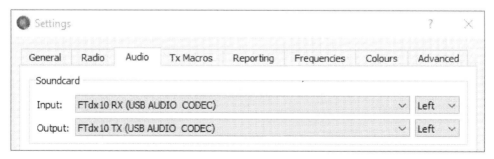

Figure 9: WSJT-X Audio settings

> ➤ **JS8Call**

The setup for 'JS8 call' is very similar to the setup for WSJT-X although 'JS8 call' doesn't support the FTDX10 yet.

Create a configuration file for the FTDX10 by cloning or renaming the 'default' setting. Select your new Config file and then click 'File' then 'Radio.'

On the 'CAT Control tab,' select FTDX101D, your Enhanced COM port, default, default, default.

On the 'Rig Options' tab Set the PTT method to RTS and your Standard COM port. In 'Mode,' select 'Data/Pkt.'

In 'Split Operation,' select Rig.

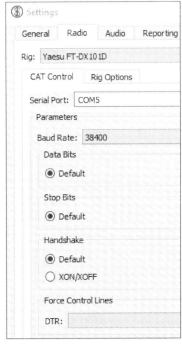

Figure 10: JS8Call CAT setup

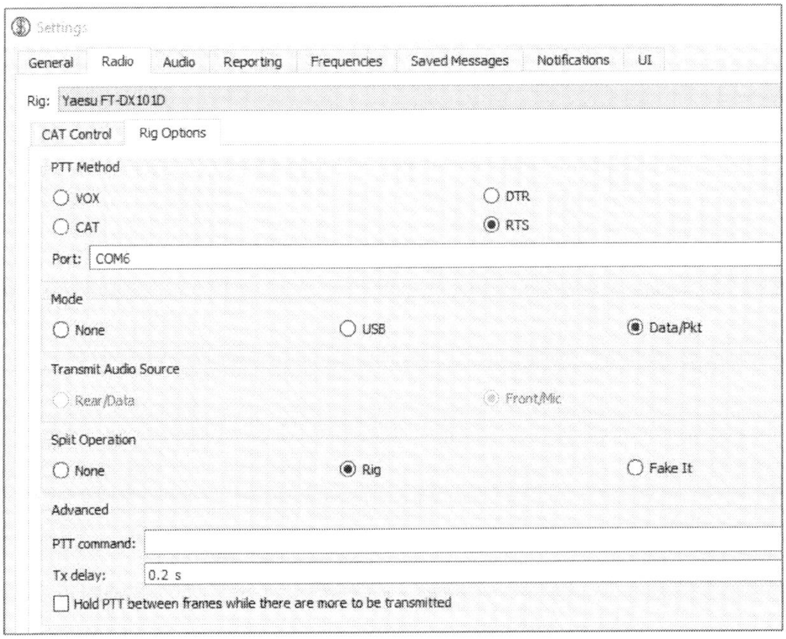

Figure 11: JS8Call RIG Options setup

➢ **MMTTY**

MMTTY can be used as a standalone RTTY program and it is often used in conjunction with logging software like DX4Win, or contest logging software such as N1MM.

MMTTY COM port settings

I had a great deal of difficulty getting MMTTY to work with the FTDX10. The answer is to use an addon called EXTFSK v1.06. MMTTY does not use any CAT control, just PTT, the RTTY data signal, and the sound card codec. So, you do not need the Enhanced COM port, just the Standard COM port. Download the correct version of EXTFSK from https://hamsoft.ca/pages/mmtty/ext-fsk.php. At the bottom of the webpage, there is a download link. The download contains some source code files which you **do not need**. Copy only the Extfsk.dll file into the folder containing your MMTTY program files. That is all you need to do. Then delete ExtFSK106.zip and any remaining extracted files.

Start MMTTY and click the 'Option (O)' tab then select 'Setup MMTTY (O).' Select the TX tab. Click 'Radio Command' and set Port to None. Then click OK.

Set the 'PTT and FSK' dropdown list to EXTFSK (not EXTFSK64). A small popup should appear. It will stay up all the time that MMTTY is working. Note that it has a tendency to hide under other open windows including the MMTTY window. Select your 'Standard' (not enhanced) COM port, FSK = DTR and PTT = RTS. (This must match the settings in the radio).

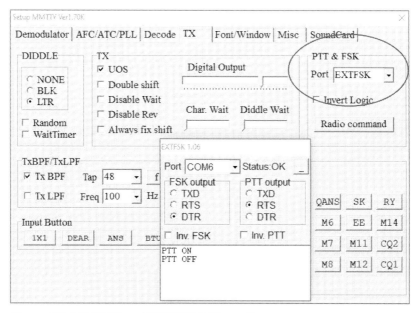

Figure 12: MMTTY and EXTFSK 1.06 config

MMTTY AFSK/FSK settings

Click the 'Option (O)' tab then select 'Setup MMTTY (O).' Select the Misc tab. For the (preferred) FSK keying mode, set 'Tx Port' to 'COM-TxD(FSK).' For the AFSK mode select 'Sound' instead. The middle option does both, which I found confusing.

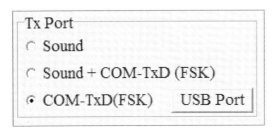

Figure 13: MMTTY FSK setting

MMTTY audio settings

Click the 'Option (O)' tab then select 'Setup MMTTY (O).' Select the SoundCard tab.

Set 'Reception' to the FTdx101D RX (USB AUDIO CODEC). If you plan to use the AFSK keying mode, set 'Transmission' to FTdx101D TX (USB AUDIO CODEC).

Reception	Transmission
⦿ FTdx10 RX (USB AUDIO CODEC)	○ Speakers (Realtek High Definiti
○ VoiceMeeter Output (VB-Audio Vo	○ VoiceMeeter Input (VB-Audio Voi
○ VAC 1 (Virtual Audio Cable)	○ Realtek Digital Output (Realtek
○ IC-705 output (USB Audio CODEC	○ IC-705 Input (USB Audio CODEC)
○ VAC 2 (Virtual Audio Cable)	⦿ FTdx10 TX (USB AUDIO CODEC)
○	○ VAC 1 (Virtual Audio Cable)
○	○ VAC 2 (Virtual Audio Cable)

Figure 14: MMTTY audio selection

Using MMTTY

Once the COM port has been set up, the EXTFSK popup box will come up every time you start MMTTY and will stay onscreen until you minimize it. It will show the CAT commands as they are sent to the radio. You can minimize it using the button beside 'Status OK,' but you can't close it. It will exit with MMTTY when you shut that down.

If you are using FSK keying, the transceiver must be in RTTY-L mode. You will be able to use the onboard RTTY message memories and the internal decoder in parallel with MMTTY. But since MMTTY has 16 message memories and its own decoder that "ain't much help."

If you are using AFSK keying, the transceiver must be in DATA-U mode. The same as when you are using other software.

➢ **MixW**

No version of MixW will work with Yaesu firmware earlier than V01-07, so make sure that the transceiver's firmware is up to date. I have successfully tested MixW 3.1, MixW 3.2.105, and MixW 4 v1.3.0.

MixW 4 has a dropdown selection for the FTDX101D. The CAT settings are selected using the 'Radar screen icon' to the left of the waterfall display. If the icon is not visible click the double arrow icon at the bottom left of the waterfall and set the CAT up as per the port settings on the image.

Port = Enhanced Com port, Parity = None, Baud Rate = 38,400, Stop Bits =2, Data bites = 8, RTS = OFF, DTR = OFF, no hardware flow control.

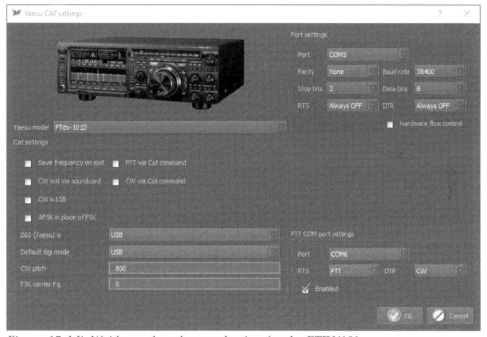

Figure 15: MixW 4 has a dropdown selection for the FTDX101

The Soundcard settings are selected by clicking the Setup 'cogs icon' at the bottom of the screen, then 'Sound Card.' Initially, I had sound input but no waterfall display. The waterfall settings are found by clicking the double arrow icon at the top right left of the waterfall. You can set the brightness, contrast, and speed.

MixW 3.1 and MixW 3.2 work well with Model set to FT-5000

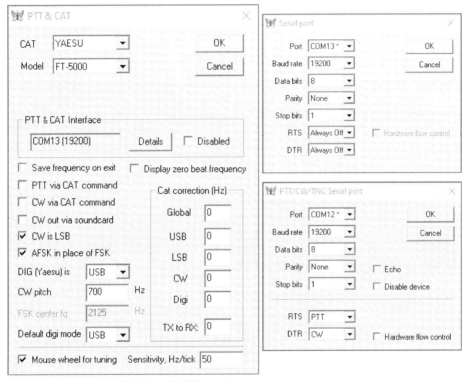

Figure 16: MixW_3 COM and PTT settings

Under 'Hardware' '**CAT Settings**' set the Yaesu and FT-5000. In the 'PTT and CAT Interface' box click on 'Details.' Set Port to the Enhanced COM port, Baud rate (same as the radio). Data Bits (8), Parity (None), Stop Bits (1), RTS (Always Off), DTR (Always Off).

Under 'Hardware' '**PTT Port Settings**,' set Port to the Standard COM port, Baud rate can be anything. I use 19200. Data Bits (8), Parity (None), Stop Bits (1), RTS (Always Off), DTR (Always Off). Data Bits (8), Parity (None), Stop Bits (1), RTS (PTT), DTR (CW).

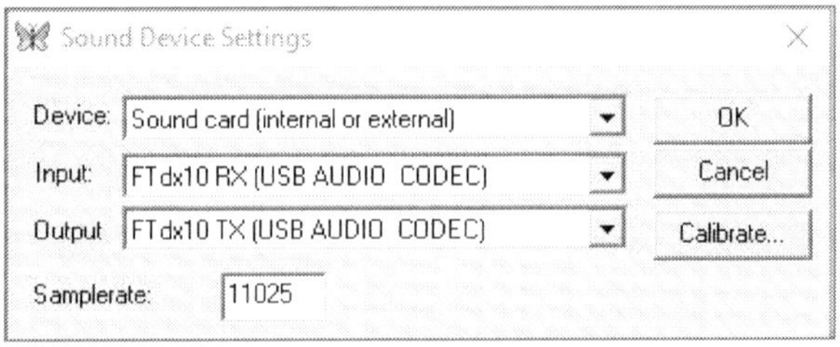

Figure 17: MixW_3 audio settings

> ➤ **Fldigi**

Fldigi usually uses HamLib or RigCAT to interface with the transceiver. But I believe the easiest way to get Fldigi to work with the FTDX10 is to download and install Flrig and use that as the radio controller.

Fldigi: Download and install the latest version of Fldigi, I am using V4.1.15. Under 'Configure' – Config Dialog – Rig Setup, check 'Enable flrig xcvr control with Fldigi as client' and 'shutdown flrig with Fldigi.' Leave everything else at the default settings. Check the CAT and Hamlib tabs to make sure that those methods of control have not been enabled.

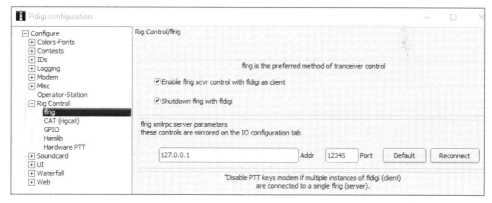

Figure 18: Fldigi setting for Flrig control

Flrig CAT control: Download and install Flrig V1.4.1 or later. Start the program with Fldigi still running. Click Config, Setup, Transceiver. Set rig to FTDX101D, Set Update to the enhanced COM port. Set the Baud rate to the setting used in the radio, 1 stop bit, and RTS/CTS signaling. I left everything else set at the default settings.

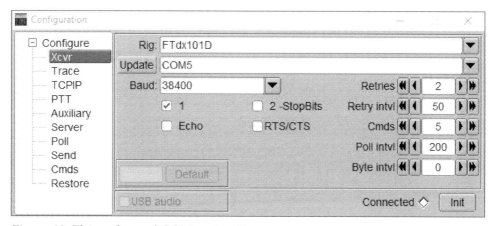

Figure 19: Flrig enhanced COM port settings

➤ **Flrig PTT control:** Click Config, Setup, 'PTT – Generic'. Select the Standard COM port and click PTT via RTS in the 'PTT control on Separate Serial Port' section.

➤ **Using Fldigi**

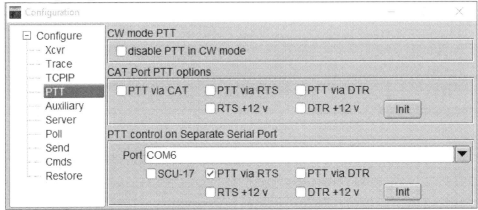

Figure 20: FLrig PTT port settings

Turn on the transceiver. Start the Fldigi program. Start the Flrig program. The frequency display on both Fldigi and Flrig should change to match the transceiver's VFO. You can tune using the radio, or Fldigi, or Flrig. There are various other controls on Flrig including volume, notch filter, I.F. shift, attenuator, preamplifier, tuner, and PTT. You can save frequencies to memory channels, change modes, and receiver bandwidth (I.F. width). You must leave the Flrig window active, for Fldigi to work but the window can be minimized.

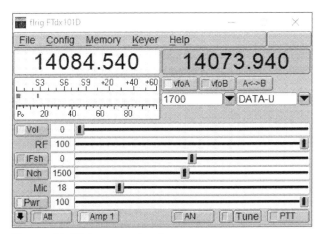

Figure 21: Flrig control software

> **Ham Radio Deluxe**

I was not able to test HRD, but the latest version of HRD supports the FTDX101, so the setup in the image should work fine. Choose the Yaesu and FTDX-101MP options, set the COM port to your Enhanced (CAT) Com port. Set the speed as per the setting in the radio. Finally, check the RTS setting but not the DTR one.

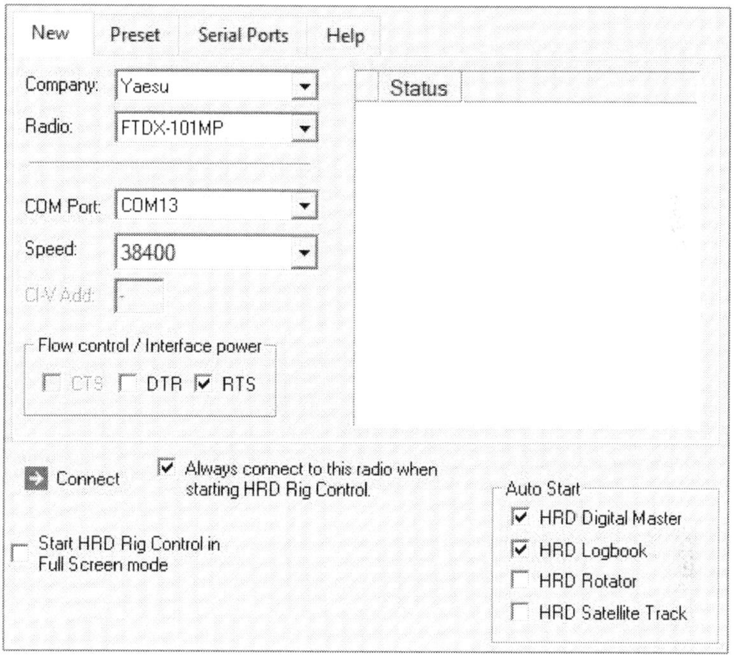

Figure 22: Ham Radio Deluxe setup

> **Digital Master (DM780)**

DM780 is usually associated with Ham Radio Deluxe. It works fine with AFSK RTTY and PSK. CW is directly keyed and that works great as well. If you enabled the Auto Start options in Ham Radio Deluxe, DMR 780 and the logging program should start at the same time as HRD.

When you use RTTY make sure that you have selected RTTY-45 AFSK.

Connecting to the Host: If DMR780 is not connecting to HRD, or there are no VFO frequency numbers displayed. Click 'Radio' to show the Radio tab. If the display says 'Closed' instead of showing the transceiver VFO frequency, there should be two buttons above the word 'Closed.' 'Connect to HRD' and 'Configure.'

The 'Configure' tab lets you choose the functions allocated to the buttons and settings below the frequency indicator. Leave the address and Port at 'localhost' and

7809 (or whatever it says). Check the Automatically connect checkbox. Click 'Save' to exit. HRD should connect and display the transceiver's frequency.

You can close the Radio Tab if you want more room on the screen.

TIP: If you want to get back to the configure menu, click the 'play' symbol above the frequency display to disconnect HRD.

The logbook tab: The logbook tab should appear beside the Radio one. If all the squares are greyed out the Logbook is not connected to HRD. Select the 'Program Options' tab and select Logbook. Leave the address and Port at 'localhost' and 7825 (or whatever it says). Check the Automatically connect checkbox. Click 'Save' to exit. HRD should connect and display white squares for log data entry.

PTT settings for CW, digital modes, or voice macros: Click the 'Program Options' tab and select PTT. This should be set for 'via Ham Radio Deluxe …'

Figure 23: DM780 PTT setting

CW setting: On the Modes and IDs tab, CW sub-tab, you can set the keying signal over the Standard COM port. Check 'Enable serial (COM) port keying' and DTR. Select the Standard COM port number. At the top of the same tab check 'Use PTT.'

RTTY setting: The FSK RTTY settings are on the RTTY sub-tab. But I was unable to get it to work. Every time I attempted to send FSK RTTY the radio would lock up and an error message said that the program could not find the Standard COM port. It finds just fine on CW mode.

Soundcard settings: Click the 'Program Options' tab and select Soundcard. Select both FTDX10 audio devices. Mine are labelled FTDX10 RX (USB AUDIO CODEC) and FTDX10 RX (USB AUDIO CODEC). It looks like you have to use the Windows level controls to set the DM780 audio levels.

➢ **Using Digital Master (DM780)**

For CW the radio should be in CW mode. This is because the DM780 software is sending CW as a digital signal via the DTR line, not as an audio tone.

For all other digital modes, including AFSK RTTY, the radio should be in DATA-U mode.

If you do manage to get FSK RTTY working, the radio would be in RTTY-L mode.

SETTING UP THE SPECTRUM AND WATERFALL DISPLAY

You will be using the touchscreen and the spectrum display all the time. This section covers the settings for the spectrum and waterfall display rather than instructions on operating the touchscreen which has a chapter all to itself.

You can use Soft Keys to change the waterfall and 3DSS spectrum colors, and the ratio of the spectrum to waterfall display. The full set of 'Display' menu settings are included on page 147 in the FUNC menu chapter. These are just the ones that affect the spectrum and waterfall display.

The spectrum display has a 50 dB dynamic range. The level can be allocated to the VFO tuning ring by press and holding the CS button and selecting LEVEL, or to the FUNC knob by selecting <FUNC> <LEVEL>. You can calibrate the spectrum scope for transmitting (page 180) but there is no separate transmit level adjustment.

➢ **Expand Soft Key**

The EXPAND Soft Key at the bottom of the waterfall display toggles between the normal spectrum and waterfall display with the ATT, AMP, R.FIL, and AGC Soft Keys and the filter function display, and the expanded display which has a larger spectrum and waterfall display.

➢ **Waterfall to spectrum display ratio**

Touch the waterfall (not the spectrum display) to change the ratio of spectrum to waterfall on the normal 2D display. There are three options. A small, medium, or large waterfall with a corresponding change in the size of the spectrum display. I prefer the 50/50 display.

➢ **Spectrum display frequency change**

Touching the spectrum display (not the waterfall) changes the frequency of the radio. It is not very precise unless you are using a narrow span, but it does provide a quick way to move the VFO near to a signal you have noticed on the scope. After that, you will probably have to fine-tune the frequency using the VFO knob.

➢ **Waterfall and 3DSS spectrum color selection**

Pressing <FUNC> <COLOR> sets 'Color' to the FUNC knob. Turn the knob to popup the 'Display Color Selection Screen' for four seconds.

Unlike the FTDX101 there is no display of the receiver bandwidth inside the roofing filter. The waterfall is the same color across the whole display.

Color choice: The colors you choose will be used for both the 3DSS and the normal spectrum and waterfall display and they affect the display on all bands.

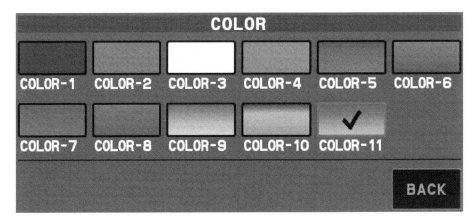

Figure 24: Color picking dialogue box

Color map: The table describes the eleven color combinations.

Color map			
Number	Waterfall/3DSS	Signal	Spectrum
COLOR-1	Blue	Blue - White	Pale blue
COLOR-2	Light blue	Blue - White	Pale blue
COLOR-3	Grey scale	Grey - White	White
COLOR-4	Amber	Orange - White	Yellow
COLOR-5	Red to Blue	Red to Blue	Pale blue
COLOR-6	Blue to Red	Blue to Red	Pale pink
COLOR-7	Violet to Blue	Violet to Blue	White
COLOR-8	Blue to Violet	Blue to Violet	Lavender
COLOR-9	Blue to Light Blue	Blue to Light Blue	White
COLOR-10	Blue	Blue to Green	Green
COLOR-11	Red to Orange	Orange	Pale yellow

➢ **Adjusting the 2D spectrum level**

Adjust the level for the 2D display until you see a line of 'grass' across the bottom of the spectrum display. With that setting, the waterfall display color should also be optimal. Note that adjusting the spectrum and waterfall display does not affect the receiver's performance at all.

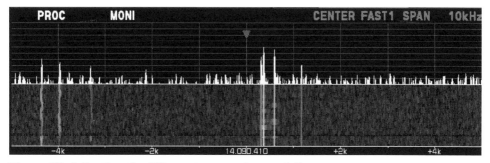

Figure 25: Adjusting the 2D spectrum and waterfall

TIP: If you adjust the grass level but you want the waterfall brighter, or less bright, you can adjust <FUNC> <PEAK>. Or you can try changing the <FUNC> <DISPLAY SETTING> <SCOPE> <2D DISP SENSITIVITY> setting to NORMAL.

➢ **Adjusting the 3DSS display level**

Adjust the level for the 3DSS display until the background noise is about a 50% mix of black and colored speckles. That way you maximize the dynamic range of the display. Signals should show as rows of spikes receding into the distance. The super-bright images on the Yaesu advertisements and most websites show the 3DSS display level set much too high.

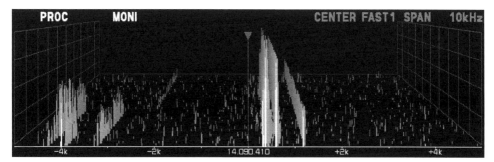

Figure 26: Adjusting the 3DSS display

TIP: If you adjust the 3DSS level but want the background brighter, or less bright, you can adjust <FUNC> <PEAK>. Or you can try changing the <FUNC> <DISPLAY SETTING> <SCOPE> <3DSS DISP SENSITIVITY> setting to NORMAL.

➢ **Spectrum scope menu settings**

RBW (Resolution bandwidth): There are three choices of resolution bandwidth for the spectrum scope. I can't see any reason why you would want to change from the default HI (high) setting. <FUNC> <DISPLAY SETTING> <SCOPE> <RBW>.

Carrier point: The easiest way to see the effect of the carrier point control is using the 'CENTER' display mode on SSB.

To change the scope center point, select <FUNC> <DISPLAY SETTING> <SCOPE> <SCOPE CTR> and select (CARRIER or FILTER).

If you choose FILTER, the receiver passband will be placed right in the center of the screen, with the white center line in the middle of the receiver passband. If you turn on the Marker you will see the real VFO frequency on the left side of the receiver passband for USB, or the right side if the radio is set to LSB.

I strongly recommend using the 'carrier point' CARRIER option which places the real VFO frequency on the white center line with the receiver passband on the right for USB, on the left for LSB, or in the center for AM, CW, or FM.

Markers: I recommend that you turn the markers on if you are using any display mode other than CENTER. A vertical line will be shown on the spectrum display at the carrier point frequency. The green marker indicates the receiver VFO frequency, and the red marker indicates the transmitter VFO frequency. In simplex mode, they overlay each other with the red marker on the top. You will be able to see both markers if you select Split operation.

To toggle the markers on or off, select <FUNC> <MARKER>

If you tune above or below the range of frequencies in the currently displayed span while using the CURSOR mode, the markers will sit at the edge of the display and the entire spectrum and waterfall or 3DSS display will scroll as you tune. Touching the spectrum or waterfall will move the VFO to a frequency close to the position you touched. On the FIX mode, the markers will sit at the edge of the display, but the display will not scroll.

2D and 3DSS sensitivity: There are 'HI' and 'NORMAL' sensitivity settings for the 3DSS or waterfall and spectrum display. Try both settings and see which you prefer.

<FUNC> <DISPLAY SETTING> <SCOPE> <2D DISP SENSITIVITY>

<FUNC> <DISPLAY SETTING> <SCOPE> <3DSS DISP SENSITIVITY>

I prefer the HI setting for both the 3DSS display and the 2D display.

Waterfall and 3DSS scroll speed: Use the SPEED Soft Key (bottom right) to toggle between the five settings. I am happy with the SLOW-2 or FAST-1 setting.

Waterfall and 3DSS peak: This setting changes the dynamic range of the waterfall display. If you are on a noisy band or there are large signals you might reduce the peak level. It also sets how bright the waterfall display will be at the spectrum "grass" or 3DSS LEVEL, you have set.

Use <FUNC> <PEAK> and rotate the FUNC knob. The selection stays associated with the FUNC knob so you can experiment and see what you like. I usually leave it set to LV2.

Spectrum level: You will be adjusting the spectrum level constantly. It changes every time you change the receiver bandwidth or band. See 'adjusting the 2D spectrum level,' and 'adjusting the 3DSS display level,' on page 55. It would be nice if the radio remembered the spectrum level for each band and had a separate level adjustment for the spectrum scope while transmitting.

The best option is to associate the spectrum level to the VFO tuning ring by press and holding the CS button and selecting LEVEL. This control is so essential I leave CS set to LEVEL and the CS function turned on all the time.

Or you can select <FUNC> <LEVEL> and rotate the FUNC knob. The selection stays associated with the FUNC knob until you make another menu selection.

CONNECTING AND USING A LINEAR AMPLIFIER

➢ **Linear amplifier connections**

Connect a PL-259 to PL-259 RF cable from the ANT output connector on the radio to the AMP INPUT connector on the linear amplifier. The linear amplifier output will be connected to the antenna, possibly via an antenna tuner and/or a Power or SWR meter.

If you have a Yaesu VL-1000 linear amplifier it will be controlled via a Yaesu CT-118 cable connected to the 'LINEAR' connector on the rear panel of the radio. This connection takes care of PTT, ALC, and automatic amplifier band switching.

If you have a non-Yaesu linear amplifier you can use the ALC and PTT connections on the 10 pin mini-DIN 'LINEAR' connector on the radio. Very unusually there are no RCA jacks for the PTT or ALC. The 'LINEAR' connector pinout is covered in 'Rear Panel Connectors,' page 170.

The TX GND output is an open collector transistor capable of sinking 200 mA at up to 60 volts, or 1 amp at up to 30 volts. It should have no problem switching even ancient linear amplifiers.

Some linear amplifiers are compatible with the Yaesu band data which is also available on the LINEAR connector. Or you could build a band decoder using an Arduino or similar. The Yaesu band data format is covered on page 170.

➢ **Linear amplifier ALC**

An ALC (automatic level control) connection between the linear amplifier and the transceiver is not essential but it is recommended. The cable goes directly from the ALC output on the amplifier to the ALC input on the LINEAR connector on the FTDX10. The ALC output from the linear amplifier turns down the transceiver output power if the amplifier is being overdriven. It should be configured so that it is not operating unless the transceiver is accidentally left at full power when driving the amplifier. Don't use it as a method of controlling the amplifier power. It should only be used as a failsafe in the event of a power setting mistake. ALC also helps to protect the linear amplifier from 'overshoot.' Some transceivers emit a full power RF spike when you key the transmitter even when the RF power is set to a low level. Without ALC control this tends to trip the protection circuit in the amplifier.

➢ **Setting the linear amplifier ALC level**

Find out how to adjust the ALC output level on your amplifier. It might be a menu setting as in the case of my Elecraft amplifier or it might be a control on the front or rear panel of the amplifier. Press <FUNC> <RF POWER> on the Yaesu radio and use the FUNC knob to turn the RF Power level down to about 30 watts. This could be higher if your linear amplifier needs a lot of drive. It is easiest to use CW with the keyer turned off, or FM, or your digital mode of choice. Do not transmit at full power for an extended period. A minute at a time should be more than enough.

Remember to identify your station as you transmit. Increase the RF POWER level until the Linear Amplifier peaks to full power, (or the highest power that you want to run). Note the RF POWER setting for future reference.

Now, while transmitting, increase the ALC level from the linear amplifier until the output power from the transceiver just starts to decrease. Then back the control off slightly so that full power is being generated and the linear amplifier ALC is not having any effect on the output power. This 'threshold point' is the correct setting. If you turn up the RF POWER control on the Yaesu transceiver, the linear amplifier should be prevented from going into an overload situation.

The ALC control is an amplifier protection method, like a circuit breaker or a fuse. You can get spurious outputs and intermodulation if you intentionally run full power from the radio and rely on the ALC to limit the RF power. Always reduce the RF LEVEL control to the safe level that you determined above. That way the ALC should never operate, but it is there in case of "finger trouble."

➢ **Digital modes**

I can't find anything in the Yaesu documentation that suggests that you have to reduce the transmitted RF power for continuous transmission modes like FM, AM, or digital modes. However, it is likely that your linear amplifier is not designed for running at full power for long periods.

Check the amplifier manual and if necessary, reduce the output for digital modes. I don't usually run the linear amplifier more than half power on digital modes. In fact, I rarely use the linear amplifier at all. It is not necessary for most contacts. 50 Watts is usually more than adequate for modes like FT8 and PSK31.

My Elecraft KPA-500 amplifier is rated for 500 Watts output for up to 10 minutes with a minimum five-minute cool-off period. But I don't find it necessary to run more than 250 Watts on digital modes.

USING AN SD CARD

The SD card is used to store the radio configuration, memory channel contents, and saved screen capture images. Or to upload new firmware. You can use a full-size SD card, but these days most people buy a micro SD card, sold with a free SD adapter. It can be a 2 Gb SD card or an SDHC card from 4 Gb up to 32 Gb. I am using a 16 Gb Kingston 'Canvas Select Plus,' micro SDHC card. The information screen says that I have 14.3 Gb free space out of the 14.4 Gb on the card, so buying a 16 Gb card, is probably overkill.

Slide the micro SD card into the adapter with the copper connector strips inserted first and the text side facing up, the same as the adapter. Then insert the SD card, or card & adapter, into the SD card slot below the transceiver touchscreen. It should be inserted into the radio with the text side facing up and the cut-off corner on the right. The card should slide in easily and click into place. It will only go in one way around. To remove the SD card, push and release it to unlatch the card (or adapter) then pull it from the card slot.

You can check the card size and amount of free space available using <FUNC> <EXTENSION SETTING> <SD CARD> <Informations DONE>.

My SD card contains files for the FTDX10 and the FTDX101D and it still has 99% free space. It would fill up more quickly if I saved off-air recordings.

➢ **Reading and writing to the SD card in your PC**

Many laptop computers feature an SD card slot, but they are not common on desktop computers. I purchased a cheap USB to SD card reader so that I can download screen shots for this book and install transceiver firmware updates from the Yaesu.com website. My USB to SD 'dongle' cost less than USD 10, but I have seen them online as cheap as 52 cents.

> ➢ **Screen capture**

You can take a screenshot picture of the display screen and save it as an 800x480 pixel 'BMP' file on the SD card. Press and hold the MODE knob until you hear a double beep and see 'Screen Shot' on the display. The files will be stored in a 'Capture' subdirectory of the SD card.

> ➢ **Formatting the SD card**

Insert an SD card into the slot below the VOX button. Yaesu says to format it before use, but I didn't bother, and everything works fine.

<FUNC> <EXTENSION SETTING> <SD CARD> <Format DONE>

> ➢ **Saving and loading memory channels**

It is a very good idea to save a backup of your memory channels onto the SD card. They may get deleted if you do a firmware update and you will need the backup to restore them. Be very careful to use the MEM LIST SAVE command. I know from personal experience that it is very easy to overwrite your memory channels with old data if you accidentally use MEM LIST LOAD.

To save a backup of your memory channel information

Use <FUNC> <EXTENSION SETTING> <SD CARD> <Mem List Save DONE> <NEW> <ENT>. Touch 'FILE SAVED' to exit.

To restore memory channel information from a backup

Use <FUNC> <EXTENSION SETTING> <SD CARD> <Mem List Load DONE>. Select the newest file and then click OK when asked to 'Overwrite.' This is your last chance to back out. The radio will produce four beeps and reboot after you touch the 'FILE LOADED' icon.

> ➢ **Saving and loading the radio configuration**

It is also a very good idea to use the SD card to save a backup of your FUNC menu settings. Your customized settings may be deleted if you do a firmware update. You will need the backup to restore them.

To save a backup of your menu settings

Use <FUNC> <EXTENSION SETTING> <SD CARD> <Menu Save DONE> <NEW> <ENT>. Touch 'FILE SAVED' to exit.

To restore your Menu settings from a backup

Use <FUNC> <EXTENSION SETTING> <SD CARD> <Menu Load DONE>. Select the newest file and then click OK when asked to 'Overwrite.' This is your last chance to back out. The radio will produce four beeps and reboot after you touch the 'FILE LOADED' icon.

> **Information**

<FUNC> <EXTENSION SETTING> <SD CARD> <Informations DONE> pops up a screen that shows the storage size of your SD card and how much space is available.

> **Installing an SD card when the radio is running**

If you install an SD card when the radio is on and not displaying a menu, you will be presented with a small popup window that says 'SETUP? YES NO.' It is quite safe to touch the YES option. It just opens the <FUNC> <EXTENSION SETTING> <SD CARD> menu. If you don't want to make any changes, touch the NO option.

FIRMWARE UPDATES

Updating the firmware is always a bit scary because if it goes wrong, it can turn your radio into a "brick." However, I found that updating the firmware went very smoothly. Some of the features discussed in this book require the firmware to be version V01-08 or newer. Firmware updates since the radio was released introduced the 'Preset' mode, changes to the band-stack, and optimization of the CW waveform rise and fall times.

It is very important to update the radio to the latest firmware release.

Do not turn the radio off during a firmware update.

> **Checking the installed firmware revision**

You can determine the currently installed firmware revisions using,

<FUNC> <EXTENSION SETTING> <SOFT VERSION>

Your firmware should be,

- Main CPU   (V01-08 or newer)
- Display CPU   (V01-03 or newer)
- Main DSP   (V01-01 or newer)
- SDR (FPGA)   (V02-00 or newer)
- AF DSP   (V01-00 or newer)

> **Downloading firmware from the Yaesu website**

Visit the Yaesu website at http://www.Yaesu.com. Find the FTDX10 section under products or by clicking the FTDX10 image on the banner.

Select the 'Files' tab. Look in the 'Amateur Radio \ Software' section for the 'FTDX10_Firmware_Update Information' file. Open the file and compare the firmware version numbers against the firmware currently loaded on the radio.

If the latest release is newer than the one currently installed on the radio, you should download the new firmware. If the latest release is not newer than the firmware already on the radio. Sit back and relax. You do not need to do a firmware update.

Download the 'FTDX10_Firmware_Update_[*revision*] (*date*),' file. Clicking the link will download a zip file to your PC. The most recent firmware included in this book is, 'FTDX10_Firmware update_202105.zip.'

Unzip the files and save them directly to the FTDX10 directory on the SD card. Do not save them into a sub-directory. If your PC does not have an SD card slot, you can buy a USB SD card reader very cheaply online or from a computer store.

After copying the files, insert the SD card back into the SD card slot under the touchscreen on the radio.

Name
Capture
MemList
Menu
MESSAGE
PlayList
Firmware_Ver_Up_Manual_ENG_FTDX10_...
FTDX10_AFCPU_V0100.SFL
FTDX10_DISPLAY_V0103.SFL
FTDX10_IFDSP_V0101.SFL
FTDX10_MAIN_V0108.SFL
FTDX10_SDR_V0100.SFL

> **Before you update the firmware**

Save your current menu settings and memory channels to the SD card before you update the firmware. They may be overwritten when you update the firmware, and you will want to recover them afterward.

To save a backup of your menu settings use, <FUNC> <EXTENSION SETTING> <SD CARD> <Menu Save DONE> <NEW> <ENT>. Touch the File Saved icon when the save has been completed.

To save a backup of your memory channel information use, <FUNC> <EXTENSION SETTING> <SD CARD> <Mem List Save DONE> <NEW> <ENT>. Touch the File Saved icon when the save has been completed.

> **Updating the firmware**

Once you have saved the backup files, downloaded the firmware file from Yaesu, saved it on the SD card, and reinstalled the card in the transceiver, you are ready to install the new firmware.

<FUNC> <EXTENSION SETTING> <SD CARD> <Firmware Update DONE>, pops up a screen which shows the firmware versions on the SD card and allows you to update the radio.

The radio software will automatically select the updates that are newer than the firmware on the radio. There is not much point in updating the other items unless you have a fault on the radio. If the downloaded zip file contains no update for one of the processors, it will indicate '(No such file).'

Select the firmware that you want to update, then touch UPDATE, and then OK.

Do not turn the radio off during a firmware update.

After the new firmware has been installed, the radio will re-boot to load the revised files into the various DSP, FPGA, and CPU devices.

If an error message is shown during the firmware update, for example, a corrupt file has been detected. Touch the screen and the radio will re-boot automatically. Then download a fresh copy of the firmware from Yaesu and repeat the update process.

Operating the radio

The Yaesu FTDX10 is a joy to use. I like the feel of the VFO tuning knob. The drag is adjustable, but I like it just the way it came from the factory. Generally, the ergonomics are good although some of the buttons are close together and close to the VFO tuning ring. I have assumed that you are familiar with the basics of operating an HF transceiver. This chapter includes some differences in the way that you operate this radio compared to other models. For example, the Split mode uses both VFOs. The memory channels and scanning functions are a bit unusual as well. It also covers, using external digital mode software, computer mouse operation, and a section on operating CW which includes the excellent APF filter. The filter function display is a great feature. It can show you the audio frequency of an interfering signal and the effect of the manual notch filter, or contour control. In CW mode an optional bar graph shows you the exact tuning of a wanted CW signal.

TIP: You may be wondering why I didn't place the chapters about the touchscreen and the spectrum display before the chapter about operating the radio. I did it for two reasons. Firstly, it is more interesting to read about how you will actually use the radio, and secondly, because the discussion about the touchscreen naturally leads on to the spectrum display and that in turn to the Soft Keys and the FUNC menu, then the sub-menu options. I believe it is natural to follow that with the front panel controls and that leads on to the rear panel connectors. So overall I believe that it was better to put the Operating the Radio chapter first.

CHANGING THE MODE

The MODE button pops up the Mode selection window. You have to be quick as it only stays visible for a few seconds.

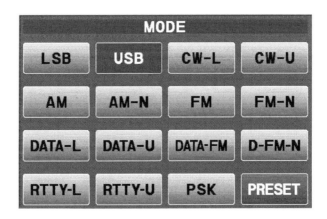

LSB is normal for SSB on bands below 10 MHz

USB is normal for SSB on bands above 10 MHz

FM is the normal mode for FM and repeaters. (FM-N would only be used if you were using the radio with a transponder).

DATA-U is the normal mode for external digital modes

DATA-FM would be used for packet radio on the 6m band

RTTY-L is the normal mode for RTTY (internal or external FSK)

PSK is for the internal PSK mode, PSK decoder, and message keyer

PRESET is active if it is blue and not active if it is grey. Touch and hold PRESET to access the Preset setup menu. Why this menu is not in the RADIO SETTING menu, beats me.

CW-L and **CW-U**. It's a personal preference, but I recommend using CW-U on bands above 10 MHz and CW-L on bands below 10 MHz. That way the CW signals behave the same way as they would if you were using single sideband. On CW-L the CW signal is on the low side of the local oscillator and tuning the VFO down increases the CW pitch. Like listening to the same signal on LSB. On CW-U the CW signal is on the high side of the local oscillator and tuning the VFO up increases the CW pitch. The same as listening to the same signal on USB. You can use either option. The VFO will indicate the received CW station's frequency and you will hear the signal at a tone equal to the CW pitch offset, <FUNC> <CW PITCH>.

➢ **Screenshot**

Press and hold the MODE button to save a screenshot image of the touchscreen display. It will be saved as an 800x480 pixel 'BMP' file on the SD card. The files are stored in a 'Capture' subdirectory of the SD card.

CHANGING THE BAND

The BAND button pops up the Band selection window. You have to be quick as it only stays visible for a few seconds.

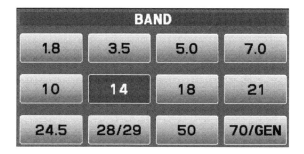

Frequency	Band	Frequency	Band
1.8 MHz	160m	18 MHz	17m
3.5 MHz	80m	21 MHz	15m
5.0 MHz	60m	24.5 MHz	12m
7.0 MHz	40m	28/29 MHz	10m
10 MHz	30m	50 MHz	6m
14 MHz	20m	70/GEN	4m & General

➢ **Band Stack register**

The FTDX10 has a band stacking register although it is a bit difficult to use. You can pre-load the band stacking register with up to three frequency and mode combinations.

Press the BAND button and select a band. There will be a slight delay and the VFO will change to the last tuned frequency on the selected band. Tune to a frequency and select the mode that you want to save in the band stacking register. Press the BAND button and select the <u>same</u> band. Tune to a second frequency and select the mode that you want to save in the band stacking register. Press the BAND button and select the <u>same</u> band. Tune to a third frequency and select the mode that you want to save in the band stacking register. Press the BAND button and select the <u>same</u> band.

Now, when you return to the band from a different band, the VFO will change to the last tuned frequency on the selected band. The third frequency you saved. Press BAND and select the same band. The VFO will change to the first of your saved frequencies. Press BAND and select the same band. The VFO will change to the second of your saved frequencies.

Note that your saved selections will be overwritten if you change frequencies and then bands. Overall, I think that it is much better to use saved memory slots or the QMB memory bank.

CS

CS stands for 'custom select.' It allows you to allocate one of sixteen different functions to the VFO tuning ring. Once allocated, pressing the CS button activates the function and you can use the VFO tuning ring to make adjustments. If you use the Split or Clarifier buttons, CS will turn off and the selected function will be allocated to the tuning ring. Pressing CS again will reactivate the custom selection.

Press and hold CS to choose one of the 16 functions. Since there is no dedicated control for adjusting the level of the spectrum scope, I usually allocate LEVEL to the tuning ring.

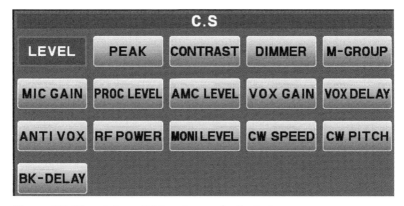

Figure 27: The sixteen CS (custom select) choices

GENERAL RECEIVER OPERATION

The FTDX10 has a fabulous receiver with some really neat features.

➢ **Tuning**

Normally for SSB, CW, or DATA, with CS turned off, the VFO knob tunes the radio in 10 Hz steps, and the VFO ring tunes in 100 Hz steps. If CS is turned on, the VFO tuning ring will adjust the assigned function instead. Usually the spectrum scope level.

In the FM and AM modes the VFO knob tunes in 100 Hz steps and the VFO ring tunes in 1 kHz steps.

The FINE tuning button decreases the VFO tuning rate by a factor of ten, to 1 Hz steps for SSB, CW, and the Data-U modes. The VFO tuning ring stays at 100 Hz steps.

You can change the tuning steps and the channel steps. See page 104 for the details.

➢ **Spectrum scope span and roofing filter bandwidth**

Use the 3 kHz roofing filter for SSB. I often select a 50 kHz span on the spectrum display. It provides a good display of the band and the representation of the bandwidth of the received signal is not too big or too small. A span of 100 kHz looks OK as well. For a contest, you may want to see more of the band so you could choose a 200 kHz or 500 kHz span. You normally use the 500 Hz roofing filter for CW. Tune onto the wanted frequency and then use ZIN or the bar graph under the S meter display for fine-tuning. Choose a span that covers most of the CW portion of the band. Anything between 20 kHz and 100 kHz looks good. I often choose a 50 kHz span because it provides a good balance between the range of displayed frequencies and the narrow receiver bandwidth. I would use a 100 kHz or 200 kHz span for FM or AM because those modes have a wider bandwidth.

➢ **Expand**

The EXPAND Soft Key creates a larger spectrum and waterfall display, overlaying the filter function display and the ATT, IPO, R.FIL, and AGC Soft Keys.

➢ **ATT, IPO, R.FIL, and AGC Soft Keys**

The 'normal' **IPO** setting is AMP1, with ATT (the attenuator) turned off. Using AMP2 is appropriate on the 10m and 6m bands where there is less band noise. And IPO can be the best option on the noisy 80m and 16m bands. Possibly with some attenuation as well if the received signals are strong. Turning off the preamplifiers by selecting IPO and adding some front-end attenuation will improve the dynamic range and as a result the signal to noise ratio of strong signals.

If you need to use the **ATT** attenuator, turn the preamplifier off (to IPO) first. Otherwise, the preamplifier is adding gain, partly negating the effect of the attenuator. You can select 6 dB, 12 dB, or 18 dB of front-end attenuation.

The receiver roofing filter, **R.FIL**, should be set to 3 kHz for the SSB and DATA-U modes, and 500 Hz for CW, PSK, or RTTY.

You can set the **AGC** to FAST, MEDIUM, or SLOW, but it is best to leave it set to AUTO and let the radio manage the AGC delay settings. If you want to change the AGC decay rates use <FUNC> <RADIO SETTING> <MODE SSB> <AGC (fast, mid or slow) DELAY>.

➢ **The 2D and 3DSS display level**

If you are using the 2D spectrum display, adjust the level so that the noise level is just showing a little 'grass' at the bottom of the spectrum display. At that level, the waterfall should have a good color, not washed out, with good contrast where there are signals. I usually allocate the Spectrum Level to the VFO tuning ring using the CS function, but you can use the FUNC control.

If you are using the 3DSS display, adjust the level so that the noise level is a mottled carpet about 50% - 80% black. I don't like it to show any 'grass,' just a newly mown lawn. That way the signals jump out with maximum dynamic range and good contrast.

➢ **RF gain and squelch**

The default setting for the RF/SQL control is for it to act as an RF Gain control. When set to 'RF,' the knob provides a manual adjustment of the gain of the RF and I.F. stages. It can be very useful to turn down the RF Gain, to improve the signal to noise ratio of reasonably strong signals, especially on the noisy lower frequency bands. But most people leave the RF Gain turned up to maximum all the time. The Yaesu manual states that the "*RF/SQ knob is normally left in the fully clockwise position.*"

I find it much more useful to change the menu setting to 'SQL' so that the control operates as a receiver squelch. If you operate mostly on 40m, 80m, and 160m, I recommend that you set the menu to the RF gain position. If like me, you favor the higher bands, set the menu to SQL.

<FUNC> <OPERATION SETTING> <GENERAL> <RF/SQL VR> (RF or SQL)

➢ **The receiver clarifier**

The receiver clarifier is equivalent to RIT (receiver incremental tuning) in earlier radios. It allows you to offset the receiver frequency without changing the transmitter frequency. The clarifier is useful if you are chatting with a station that is slightly off frequency. Rather than move the VFO which will often cause the other station to adjust its frequency the next time you transmit, you can use the receiver clarifier to fine-tune the incoming signal. It is especially useful in a net where most stations are on frequency, but one station is a little off frequency.

Pressing the CLAR RX button enables the receiver clarifier. An offset indicator will appear underneath the VFO frequency display. The clarifier offset is adjusted with the VFO tuning ring. Press and holding the CLAR RX button resets the offset to zero. Adjusting the clarifier offset will change the VFO numbers to show the offset receive frequency. The VFO will change to the transmitter frequency when you transmit.

The effect of the receiver clarifier is shown differently depending on which display type you have chosen. Turn the Markers on using <FUNC> <MARKER>. On a FIX or CURSOR display, the green marker will move away from the red marker to indicate the revised receiver frequency. The green marker is not shown on the CENTER display, because the receiver frequency remains in the center. The red transmitter marker will change to indicate the split as you adjust the receiver offset.

When you transmit, the green marker disappears and the VFO changes to the transmitter frequency. The red marker moves to the center of the display.

TIP: once the clarifier offset has been set, you can turn the clarifier on and off by pressing the CLAR RX button whenever the off-frequency station is transmitting.

➢ **I.F. Shift & Width**

The WIDTH (outer) control changes the bandwidth of the 24 kHz second I.F. inside the DSP stage of the receiver. On SSB it is adjustable from 300 Hz to 4 kHz. The default is 3 kHz to match the bandwidth of the roofing filter. On CW, PSK, RTTY, and the data modes it is adjustable from 50 Hz to 3 kHz and the default is 500 Hz to match the bandwidth of the 500 Hz roofing filter. The I.F. bandwidth is fixed at 9 kHz on AM and 16 kHz on FM.

The current DSP bandwidth is shown graphically on the filter function display below the VFO frequency display.

The white lines indicate the shape of the filter response. If you change the I.F. width the white lines move but a highlighted area remains to show the bandwidth of the roofing filter. Selecting the 'Sharp' 18 dB per octave setting does not change the displayed filter shape. An orange dip (or peak) indicates the use of the Contour control. There are also indicators for the manual notch filter, RTTY mark and space, and the CW APF filter. As you adjust the WIDTH control a popup shows the bandwidth in Hertz. It also pops up the filter function display if the radio is in the expanded display mode.

The SHIFT control moves the 24 kHz DSP filter passband by up to ±1200 Hz to give you some ability to reject interference that is inside the DSP passband, close to the wanted signal. Shifting the DSP I.F. is indicated on the filter function display. A popup displays the amount of shift you have selected as you make adjustments.

Push and holding the SHIFT knob resets both the I.F. Shift and the I.F. Width functions.

If you are still experiencing interference within the DSP bandwidth you can employ the digital notch filter, the manual notch filter, and the contour control.

> ### DNF digital notch filter

The automatic digital notch filter is effective at removing fixed carrier signals that are creating beat notes on SSB. It can remove the effect of multiple interfering carrier signals that are within the audio bandwidth. To test it, tune the radio so that you can hear a stray carrier signal. You should be able to see the signal on the filter function display and the AF-FFT display if you select MULTI.

Use <FUNC> <DNF> to turn on the digital notch filter, then return to the main display. You will hear and see that the filter has removed the offending audio signal.

TIP: you will still see the offending signal on the spectrum and waterfall display because that is showing the incoming RF signal, well before the digital notch filter in the DSP stage.

> ### NOTCH manual notch filter

Turning the NOTCH knob automatically turns on the NOTCH filter button. An orange V indicating the notch appears on the filter function display. Turning the knob changes the notch frequency and this is depicted by moving the notch across the filter function display spectrum, and in Hertz on a popup display.

Press and hold the NOTCH button to reset the notch frequency to the center of the roofing filter passband.

I normally tune the manual notch to the right spot by listening to the interfering signal disappear. Once you have adjusted the notch frequency you can toggle the filter on and off using the NOTCH button

You can change the width of the manual notch filter. The narrow setting will be fine for single carrier interference, and it has less impact on the quality of the wanted signal. The default wide setting will be better at nulling out carriers that are modulated with noise or narrowband data.

<FUNC> <OPERATION SETTING> <RX DSP> <IF NOTCH WIDTH>

TIP: if the interfering signal is extraordinarily strong, Yaesu recommends trying the manual filter before the DNF filter, because the manual notch is deeper than the auto notch.

➢ CONT

The contour control was new to me, and I quite like it. It produces a small dip or boost in the audio frequency response which you can move across the audio passband from 50 to 3200 Hz. It's like a 'mini-notch filter.' You can hear the effect as you tune across the audio signal. The Contour knob can be used as a sort of tone control, attenuating or boosting the low or high audio frequencies. This can be useful if a received signal is "rumbly" or "hissy," but you don't want to adjust the tone settings in the menu structure. It can also attenuate an interfering signal without notching out the audio frequency completely. If you set the contour to a peak, it can boost a wanted narrow band (digital or CW) signal, especially if you are using a wider roofing filter.

TIP: If you have a hearing impairment, for example, hearing loss at high frequencies you could set the contour level to a positive value and use the contour function to boost the high audio frequencies.

Turning the CONT knob (the outside ring of the NOTCH knob) automatically turns on the CONT button. An orange 'dip' or 'bump' indicator is shown on the filter function display. Once you have adjusted the contour frequency you can toggle the filter on and off using the CONT button. Press and holding the CONT/APF button resets the Contour filter to 1500 Hz. This is above the CW passband.

TIP: Contour adjustment of the higher audio frequencies has little effect on CW because the 500 Hz roofing filter attenuates those frequencies anyway.

Note: when you are using CW, the CONT/APF button cycles through APF, CONT, and OFF. CONT is shown as an orange dip, or peak, on the filter function display. You can hear the effect of the APF, and it is shown as narrowing of the spectrum on the filter function display.

You can change the bandwidth and the gain of the contour filter. You can even make it a hump so that it acts even more like a tone control, amplifying a part of the audio bandwidth. As an experiment, I set the control to +18 and you could clearly see the increase in noise level as I tuned the contour control across the audio band.

The Contour level default setting is -15. The default Contour filter width is 10.

<FUNC> <OPERATION SETTING> <RX DSP> <CONTOUR LEVEL> (-40 to +20)

<FUNC> <OPERATION SETTING> <RX DSP> <CONTOUR WIDTH> (1 to 11)

➢ **APF**

The APF (audio peak filter) button is for CW only. It is disabled on the other modes. Tune in a CW signal by ear or using ZIN or the audio bar-graph indicator and turn on APF. The very narrow audio filter eliminates nearly all band noise revealing a clean CW signal. "It is fab!"

TIP: The APF filter is very narrow. Unless you know the exact frequency, it is best to turn off APF while you are tuning across the band and turn on the APF after you have the signal tuned in.

The APF filter frequency is adjustable. Turning the CONT/APF knob while in the CW mode will activate the APF button and enable the knob. APF is adjustable over a range from -250 Hz to + 250 Hz. If the CW signal is tuned properly, you will probably not need to adjust the APF filter frequency. Press and holding the CONT/APF button resets the offset to zero.

You can also change the bandwidth of the APF filter. Using the narrow filter setting will provide the cleanest sounding CW signal but the signal has to be exactly on frequency, or you will hear nothing. The wide setting is easier to tune, but you will hear more band noise. To be honest, although I can hear a small difference, I don't think it matters which setting you choose.

<FUNC> <OPERATION SETTING> <RX DSP> <APF WIDTH>

➢ **NB noise blanker**

Pressing the NB button turns on the noise blanker. Holding the NB button down enables a popup display so that you can use the FUNC knob to adjust the noise blanker threshold level. Increasing the NB level, lowers the threshold, making the noise blanker react earlier. Setting the NB to an aggressive (high) level may affect audio quality, as it will start to trigger on voice peaks as well as noise peaks. So, you should experiment with the level when tackling a particular noise problem.

The noise blanker is designed to reduce or eliminate regular pulse-type noise such as car ignition noise. Most DSP noise blankers work by eliminating or modifying noise peaks that are above the average received signal level. They usually have no effect on noise pulses that are below the average speech level.

You can also adjust the width and the amount of the attenuation that is applied when the noise blanker reacts to noise above the noise blanker threshold.

<FUNC> <OPERATION SETTING> <GENERAL> <NB WIDTH> (1, 3, or 10 ms)

<FUNC> <OPERATION SETTING> <GENERAL> <NB REJECTION> (10, 30, 40 dB)

The NB WIDTH control should be set to a short time constant for impulse noise such as vehicle ignition or electric fence noise. You would use a wide setting for longer noise pulses such as lightning crashes or electric welding noise. Noise blankers are usually implemented early in the DSP process before any DSP filters or demodulation occurs. They "look ahead" in the digital data streams and detect signals that are of short duration and significantly larger than the average speech level. The noise blanker algorithm either eliminates or attenuates the noise pulse.

In some radios the DSP noise blanker mutes the noise pulses completely, other developers believe that the receiver sounds more natural if the level in the data sample is reduced to the average of the speech level or attenuated by a fixed value. In this radio, you have a choice of how much NB REJECTION attenuation is applied to data samples where a noise spike has been detected. The default is 30 dB, but you can select a less aggressive 10 dB or a more aggressive 40 dB.

TIP: at the risk of being flamed for repeating myself, here are some more thoughts about the NB LEVEL. If you use a high NB level (low threshold) the noise blanker will see speech peaks as noise and will attenuate them making the received audio sound distorted. On the other hand, a low NB level (high threshold) setting may mean that the noise blanker fails to react to noise pulses that are only just above the average speech level. It's a trade-off. You will have to make a judgment call on the best setting for a particular problem. Sometimes slightly distorted audio is better than putting up with distracting noise pulses. Start with a low threshold level and increase it until you reach an optimum setting.

➤ DNR digital noise reduction

Digital noise reduction is mostly aimed at beating atmospheric wideband noise. It is more useful on the noisy lower bands. It appears that the DNR gets more aggressive as you turn the control down from 15. The Yaesu manual states that there are "15 different noise-reduction algorithms" which suggests that the control is not just a linear progression and that some of the intermediate settings may work better than the higher or lower options.

Press and holding DNR pops up the adjustment window. Use the FUNC knob to change the DNR algorithm setting.

Aggressive levels of digital noise reduction can add an electronic synthesizer 'Dalek' sound to the speech but also works harder to eliminate the background noise. Which do you prefer, and which is less tiring to listen to?

TIP: Digital noise reduction takes quite a long time to calculate the noise characteristic. It can take a second or two to settle into a stable state.

The DNR works very well. Better than in the FTDX101 in my opinion. High settings eliminate all of the background noise. I found settings 5 and 9 to be very good.

> ➤ **DNF, NOTCH, NB, and DNR compared**

The NB, DNR, NOTCH, and DNF functions form your DSP toolkit for managing different types of interfering signals. The **digital notch filter** (DNF) works well with fixed signal interference such as carrier signals or ADSL interference. The filter looks at the time constant of the incoming signal and rejects signals that have a long time constant compared to the ever-changing speech. It is useless for combatting impulse noise but great for eliminating stray carrier signals known as "birdies." The DNF can respond to multiple interference signals within the receiver passband. You cannot use it when you are receiving CW because it will filter out the CW signal that you are listening to. The **manual NOTCH filter** is deeper than the DNF auto-notch filter and gives you control of where the notch is placed on the audio spectrum.

Digital noise reduction (DNR) works to reduce atmospheric and other long-term background noise levels, particularly on the noisy lower bands, 40m, 80m, and 160m. It rejects signals with a long time constant relative to normal speech or CW signals. The **noise blanker (NB)** uses similar DSP code, but different time constant variables. This time the short time constant pulse noise signals are rejected, and the longer-term coherent speech is retained. The noise blanker is designed to reduce or eliminate regular pulse-type noise such as car ignition or electric fence noise. The DSP process looks ahead in the data stream and attenuates the noise spikes with a pre-set (adjustable) amount of attenuation.

> ➤ **Markers**

The green marker indicates the receiver frequency, and the red marker indicates the transmitter frequency. The green receiver marker is turned off if the spectrum display is set to CENTER since the receiver frequency is always in the center of the spectrum display. During normal simplex operation, the red marker sits on top of the green marker, so you can only see the green marker if you are operating in the FIX or CURSOR screen mode with Split, or a Clarifier enabled.

Red or green arrows at the ⊗ left or right ⊗ side of the spectrum scope indicate that the transmitter or receiver frequency is off the edge of the currently displayed spectrum display. This does not affect the operation of the radio, but it means you can't see the relevant frequencies. I recommend leaving the markers turned on all the time unless you always use the CENTER display, and you never use split. <FUNC> <MARKER> toggles the markers on and off.

TRANSMITTING

There aren't many mysteries when it comes to operating the transmitter. Assuming that you have set all the audio levels and CW settings, as discussed in the earlier sections. Everything is much the same as using any other HF transmitter.

Other than the mode controls which affect both receiving and transmitting, only eight controls directly affect transmitter operation. They are BK-IN, TUNE, MOX, VOX, MONI, CS, the Speech Processor, and the transmitter clarifier (CLAR TX).

The biggest risk is inadvertently transmitting on the wrong band or mode. You must be careful because this is fairly easy to do.

➢ **Break-in**

BK-IN must be enabled to send CW or the five voice keyer recordings. If <FUNC> <BK-IN> is not turned on, you will hear the message or the CW sidetone, but it will not be transmitted.

You can select SEMI or FULL break-in using <FUNC> <CW SETTING> <MODE CW> <CW BK-IN TYPE>. There is some relay clicking when you use the full break-in mode, but it is not too objectionable. I recommend that you use semi break-in unless you are proficient with QSK full break-in operation.

➢ **Antenna tuner**

Pressing the TUNE button activates the radio's internal tuner. It will automatically select the capacitor and inductor that will provide a near match at the current operating frequency. The radio stores the tuner settings after tuning and retrieves them the next time you use a frequency within the same 10 kHz window.

The antenna tuner will not automatically initiate a 'tune' operation if the SWR is high. You have to manually start a 'tune' by press and holding the TUNE button.

TIP: be careful when using the tuner for the first time on a band. I selected TUNE but did not carry out a manual tune operation. The SWR was showing as infinite with a HI-SWR alarm, even though the antenna is resonant on the chosen frequency. If you are on a previously unused frequency, make sure that you initiate a 'tune' operation before attempting to operate with the tuner.

The tuner can match load impedances between 16.5 ohms and 150 ohms, (up to 3:1 SWR) on the HF bands, or from 25 to 100 ohms (up to 2:1 SWR) on the 6m band. This means that it should be able to tune out to the band edges of 'resonant' antennas, but it might not be able to tune antennas like G5RV, large loops, long wire, or 42 m verticals. This performance is typical of the antenna tuners in most amateur radio HF transceivers.

Holding down the TUNE button initiates an automatic tuning sequence. You will hear all manner of clicks until the radio arrives at the best tuning position. It seems that holding down the TUNE button always initiates a full tune, not starting from any previously saved setting.

After tuning, the tuner will save the tuner settings if the tuned SWR less is than 2:1. A HI SWR icon will be displayed if the final SWR is greater than 3:1.

➢ **External antenna tuner**

The 8 pin TUNER jack on the rear panel is for connection to the Yaesu FC-40 external tuner or a compatible alternative. The pinout for the 8 pin TUNER jack is in the Yaesu manual (page 13) but you will use the cable supplied with the tuner.

If you have the Yaesu FC-40 tuner, you must tell the radio that an external Yaesu compatible antenna tuner is connected by setting 'Tuner Select' to EXT.

<FUNC> <OPERATION SETTING> <GENERAL> <TUNER SELECT> <EXT>

When a Yaesu tuner is connected, the internal antenna tuner will be bypassed, and the TUNE button will control the FC-40 tuner instead. If the FC-40 tuner is not connected or a non-Yaesu tuner is connected, the internal tuner will operate normally.

TIP: you should not attempt to use the internal tuner and a non-Yaesu external antenna tuner at the same time. Turn off the internal tuner with the TUNE button and use the external tuner.

Note: Yaesu states that a ferrite isolator must be placed at the transceiver end of the tuner control cable for the FC-40 tuner. This is to stop RF energy from entering the radio through the tuner interface.

The ATAS-120A antenna does not have a connection to the Tuner jack. It receives control (tuning) information over the coax feeder cable. Set 'Tuner Select' to ATAS.

<FUNC> <OPERATION SETTING> <GENERAL> <TUNER SELECT> <ATAS>

When the ATAS-120A antenna is connected, the internal antenna tuner will be bypassed, and the TUNE button will control the antenna tuning instead.

➢ **VOX**

VOX stands for 'voice operated switch.' It is available for the speech and data modes. When VOX is turned on, the radio will transmit when you talk into the microphone without you having to press the PTT button. The function is popular if you are using a headset or a desk microphone. Some people always use it. I have never used it.

Turn VOX on using the VOX/MOX button. An amber LED indicates VOX is active.

The three VOX control settings are available directly from the FUNC menu. Turn the FUNC knob to change the setting. Or they can be allocated to the VFO tuning ring by press and holding the CS button. You should take the time to set the VOX up carefully as some settings tend to counteract other settings.

- <FUNC> <VOX GAIN> sets the sensitivity of the VOX, i.e. how loud you have to talk to put the radio into transmit mode. (Default 10).

- <FUNC> <VOX DELAY> sets the pause before the radio reverts to receive mode. It needs to be set so that the radio keeps transmitting while you are talking normally but returns to receiving in a reasonable time after you have finished talking. (Default 100 milliseconds).

- <FUNC> <ANTI-VOX> stops the VOX triggering on miscellaneous noise like audio from the speaker or background noise. Higher values make the VOX less likely to trigger. The default is zero i.e., no anti-VOX.

TIP: You can get a feedback loop where VOX is triggered by speech from the radio's speaker if you are using the transmit monitor, MONI. The easiest solution is to turn off the transmit monitor. Or you could try turning down the monitor volume and increasing the anti-VOX a bit. Better still, wear headphones while transmitting so that there is no feedback path to the microphone.

The 4ᵗʰ dimension. You can use VOX with digital mode transmissions rather than messing around with the RTS and DTR control line signals. I would only rely on this method of creating a PTT signal as a last resort if you cannot get standard RTS/DTR COM port signaling to work.

Change <FUNC> <OPERATION SETTING> <TX GENERAL> to DATA

There is a fourth VOX setting deep in the menu structure that sets the threshold level for 'Data VOX.' The range is 0 to 100 and the default is 50.

<FUNC> <OPERATION SETTING> <TX GENERAL> <DATA VOX GAIN>

➤ MOX

MOX stands for 'manually operated switch.' It makes the transceiver switch to transmitting in the same way as pressing the mic switch. The radio will keep transmitting until you turn the MOX off. Press and hold the VOX/MOX button to activate the MOX. The usual red TX LED indicates that the radio is transmitting, but the orange VOX LED stays off.

Note that in CW mode MOX will allow you to send Morse code from a CW message, or the Morse key or paddle, without having BK-IN turned on. The microphone is live in the voice modes, so you will transmit any conversations, phone calls, talking to the cat, praising the dog, background music, or audio from another radio. The voice keyer will operate while the radio is being keyed via MOX, but as usual, BK-IN must be turned on.

➤ MONI

<FUNC> <MONI> is the transmit monitor. It allows you to hear the signal that you are transmitting. If you are operating in SSB or AM mode, it is easier to hear and less distracting if you use headphones. You will hear the voice keyer audio, even if the transmit monitor is turned off. The transmit monitor is not available in FM.

The audio for the transmit monitor is taken from the transmitter I.F. after the modulator but before up-conversion to the final transmit frequency. This makes it especially useful for evaluating the quality and tone of the signal that you are transmitting. Use it when you adjust the parametric equalizer and speech processor (compressor) settings.

You must have MONI turned on to hear the CW sidetone. Changing the CW sidetone level does not affect the transmit monitor level in the voice modes, and vice versa.

Selecting <FUNC> <MONI> allocates the monitor (or sidetone) level to the FUNC knob. The monitor level is indicated on a popup, but it is better to just adjust it until you are happy with the volume. Alternatively, you can press and hold CS to allocate the monitor level to the VFO ring.

➢ **The transmitter clarifier**

The transmitter clarifier is equivalent to XIT in earlier radios. It allows you to offset the transmitter frequency from the receiver frequency. It can be another way of operating Split without using the split button. Press CLAR TX and use the VFO tuning ring to adjust the transmitter offset up to 9990 Hz above or below the receiver frequency.

If you are using the CENTER display with the markers turned on, you will see the red transmit frequency marker move away from the center as the offset is increased. In the FIX and CURSOR modes, you will see the red transmit frequency marker move away from the green receiver frequency marker as the offset is increased. Press and hold the CLAR TX button to reset the clarifier offset to zero.

➢ **The speech processor**

The speech processor is only available in SSB. It increases the average power of your transmission by amplifying the quieter sounds while ensuring that louder sounds do not overmodulate the transmitter. In other words, it reduces (or 'compresses') the dynamic range of the modulating audio. The result is more talk power but less natural-sounding speech.

Most people do not use speech compression for local contacts or nets because the quality of the transmitted audio is important, and your signal is usually strong. You probably would use the speech processor if you were working a contest or trying to break a 'pile-up' to work a DX station. In that situation, the higher average power helps to cut through the noise of the pile-up, which outweighs the need for high-fidelity audio. That said, too much compression can make your voice sound distorted and hard to copy, which will actually reduce your chances of working the rare DX.

Unusually you cannot just press a button or touch a Soft Key to turn on the speech compressor. You also have to turn up the compression level. Allocate PROC LEVEL to the CS button, or press <FUNC> <PROC LEVEL> to allocate the processor level to the FUNC knob. Then set the level to the number you discovered during the SSB setup on page 11, or for voice peaks around 10 dB on the COMP meter while transmitting. When the speech processor is active, a white PROC indicator is displayed at the top of the spectrum display.

You can monitor the effect of introducing the speech compressor by listening to your transmitted signal using the transmit monitor (MONI), or visually by enabling the audio scope and spectrum display using the MULTI Soft Key.

TIP: It is much better to use headphones to monitor your transmitted SSB signal.

OPERATING IN SSB MODE

By now you will have adjusted the microphone gain, the AMC gain, and the speech processor (compressor) levels. And we have covered the voice message keyer, VOX, MOX, transmit monitor, receiver clarifier, and transmitter clarifier. It is time to listen to some signals and maybe get brave enough to call CQ.

➢ **Select SSB**

Select lower or upper sideband mode using the MODE button.

Select the CURSOR or CENTER display mode on the 2D spectrum display, or the 3DSS display. If the markers are turned on, you should see a red marker. Tune in an SSB signal by ear or by tuning until the marker is at the left edge of a USB signal on the panadapter or the right side of an LSB signal. Hard to explain, easy to do.

The receiver controls and filters have been covered already. It is normal to set the FTDX10 transmitter for 100 Watts. You would normally use the speech processor if you were working DX stations or in a contest. You probably would not use it if you are chatting to a friend on a Net.

A green 'receive frequency' marker indicates that the radio is receiving on a different frequency to the transmitter. (Split or Clarifier is operating).

➢ **Transmitting**

The AMC and microphone levels have already been set so you can just push the microphone PTT and make your call.

If you want to use the Speech Processor, turn it on by touching <FUNC> <PROC LEVEL> and adjusting the FUNC knob for a level that results in less than 10 dB of compression on the COM meter. If in doubt set it to 10. You can allocate the PROC LEVEL to the VFO tuning ring using the CS button, but I wouldn't bother.

➢ **Soft Keys**

ATT = OFF, IPO = AMP1 (usually), R.FIL = 3 kHz, AGC = AUTO.

➢ **Span**

I usually use a 50 kHz or 100 kHz spectrum span when using SSB.

➢ **Voice keyer**

You can key the voice messages from the <FUNC> <MESSAGE> menu popup, or the FH-2 external keypad. Touch message 1 to 5 to send the pre-recorded voice message. You must turn BK-IN on to transmit a voice message on the air. Otherwise, you will hear it, but nobody else will. The message will only play once. There is no 'auto-repeat' function on the voice keyer.

To stop a message during transmission, touch the message number icon again.

The MEM Soft key is used to record the messages. This was covered back on page 16. RX LEVEL, adjusted with the FUNC knob, sets the playback volume of the messages. I have it set at 26%. TX LEVEL sets the transmitted level of the messages. It should be set for ALC and COMP meter readings that are similar to when you are using the microphone. I have it set to 60%.

OPERATING IN CW MODE

➢ **Tuning CW**

Tune the VFO so that the marker or center marker is exactly on the CW signal. Use the ZIN/SPOT button to set the frequency exactly. Then you should turn on the amazing APF filter. Most of the background noise will disappear. CW tuning is also indicated on the bar meter below the S meter. When the CW signal is tuned correctly, there will be three white dots in the center of the meter display with the center dot below the red indicator on the bar graph.

➢ **ZIN/SPOT**

ZIN is an auto-tune feature that pulls the receiver frequency onto a CW signal by matching the CW note with the Keyer pitch, netting it to the transmit frequency.

While receiving a CW signal, press ZIN/SPOT to pull the receiver VFO so that the received tone is the same as the keyer pitch. At that point, the transmitter will send on the same frequency as you are receiving.

Most of the time it works very well, although you may have to press the button two or three times to get the receiver exactly on frequency.

If you press and hold the ZIN/SPOT button, the keyer pitch tone will be heard, so you can compare the tone of the incoming signal with the keyer pitch. You can tune the VFO and 'zero beat' the audio from the two sources.

➢ **APF filter**

The APF button is only active in the CW mode. It enables the 'Audio Peak Filter' which is very effective at lifting weak CW signals out of the noise or eliminating other CW signals that are very close to the receiving frequency. It is best used in conjunction with the ZIN/SPOT button. Use ZIN to pull the radio onto the exact frequency being used by the transmitting station and then use APF to eliminate everything except the wanted station.

When the filter is engaged an orange line indicating the APF frequency is displayed on the filter function display.

TIP: if you see an orange dip instead of a line, you have selected the Contour filter instead of APF. Press the CONT/APF button again to turn that off, and again to turn the AFF on.

The APF filter frequency is adjustable over a range from -250 Hz to + 250 Hz. Turning the CONT/APF knob while in the CW mode will activate the APF button and a popup will display the filter offset in Hertz. Press and hold the CONT/APF button to reset the offset to zero.

TIP: unless you are on a Net with multiple CW operators, I think it is more likely you will leave the APF at the default center frequency and adjust the radio VFO so that the CW signal is centered on the filter.

You can also change the bandwidth of the APF filter. Using the narrow filter setting will provide the cleanest sounding CW signal but the signal has to be exactly on frequency, or you will hear nothing. The medium and wide settings are easier to tune, but you will hear more band noise. To be honest, although I can hear a small difference, I don't think it matters which setting you choose. <FUNC> <OPERATION SETTING> <RX DSP> <APF WIDTH>

➢ **Soft Keys**

ATT = OFF, IPO = AMP1 (usually), R.FIL = 500 Hz, AGC = AUTO.

➢ **Span**

I usually use a 20 kHz or 50 kHz spectrum span for CW.

➢ **Break-in setting**

The transceiver will not automatically transmit CW unless BK-IN has been selected using <FUNC> <BK-IN>. The radio will operate in full Break-in or Semi break-in depending on the menu setting.

<FUNC> <CW SETTING> <MODE CW> <CW BK-IN TYPE> (SEMI or FULL)

If BK-IN is set to OFF the transceiver can be made to transmit by pressing the MOX button, pressing the PTT button on the microphone, sending a CAT command, or grounding the PTT line on the RTTY/DATA jack.

The break-in setting affects the sending of keying macro messages and Morse Code send from a key or paddle, but not CW sent from an external computer program.

- With BK-IN OFF you can practice CW by listening to the side-tone without transmitting.

- Full break-in mode BK-IN FULL will key the transmitter when each CW character is being sent and will return to receive as soon as the key is released. This allows for reception of a signal between CW characters.

- Semi break-in mode BK-IN SEMI will key the transmitter while the CW is being sent and will return to receive after a delay when the key is released. The Semi Break-In delay is set using <FUNC> <BK-DELAY>. It is adjustable from 30 ms to 3 seconds. The default is 200 ms. This mode is preferred because it avoids unnecessary PTT relay switching.

➢ **Sidetone**

The CW sidetone is only heard if the transmit monitor (MONI) is turned on. Select <FUNC> <MONI LEVEL> and adjust the monitor level with the FUNC knob. This adjustment does not affect the transmit monitor level on SSB.

➢ **Key speed and pitch**

The <FUNC> <CW SPEED> control sets the speed of the electronic keyer including CW sent from the message memories. As you adjust the control the CW speed in WPM (words per minute) is displayed on the usual popup. The <FUNC> <KEYER> control turns the electronic keyer on or off.

<FUNC> <CW PITCH> changes the pitch (tone) of a received CW signal without changing the receiver frequency. You can set the control so that CW sounds right to you. I use 750 Hz. Naturally, it also sets the sidetone frequency of your transmitted signal. The CW pitch is displayed on the usual popup window, as you adjust the pitch tone using the FUNC knob.

➢ **CW keyer messages**

There are five CW messages which can be used for DX or Contest operation or just to save you sending the same information over and over. They are great for sending CQ on a quiet band. To use them, select CW mode and then <FUNC> <MESSAGE>.

If you have the external FH-2 keypad plugged into the REM jack, you can send the five CW messages by pressing buttons 1 to 5. As always, BK-IN must be on for the CW signal to be transmitted.

Provided at least one CW Memory slot is set to MESSAGE, you can press the MEM button to record a message from the paddle into that slot. This is useful for short-term storage, for example, to hold the other station's callsign.

If you have inserted the # character after the signal report into any of the TEXT messages. The 'contest' number will automatically increment each time the message is sent. If you need to send the same number again, you can decrement the number with the DEC Soft Key in the MESSAGE window, or by pressing the DEC button on the FH-2 keypad. The message keyer setup instructions are on page 17.

➤ **CW decoder screen**

I was very excited about the CW decoder because I am hopeless at Morse code. It works reasonably well although for best decoding results you need to set the speed of the CW keyer to something close to the speed of the code that is being received. Use <FUNC> <DECODE> to turn it on.

I have a few grumbles which are common to all of the decoder screens.

1. There are unused Soft Key buttons on the Decoder screen that should be used for the memory keyer messages but are not.

2. There is no 'Clear Screen' function.

3. You cannot use an external keyboard to type messages.

4. You cannot see the spectrum display when the decode screen is active, or switch between the two screens easily. If you close the decode window to see the spectrum display, you have to go back through the FUNC menu to get it back again.

5. The message sending popup blocks the received text, so you can't see the message that you are replying to. Also, it disappears if you touch almost any other control.

OPERATING SPLIT

Working in the 'Split' mode is a very common requirement if you are trying to work a rare DX station or DXpedition when they have a 'pileup' of stations calling them.

To work 'Split' you usually set VFO-A to receive the frequency that the DX station is using and transmit on the frequency indicated by VFO-B. On bands above 10 MHz, you transmit USB on a frequency that is a few kHz higher than the DX station. For bands lower than 10 MHz, you transmit LSB on a frequency that is a few kHz lower than the DX station.

The operation is the same for CW, but the split offset is 1-2 kHz rather than the 5-10 kHz offset used for SSB. If you are the rare DX or DXpedition station, you would, of course, reverse the split.

You can operate Split on digital modes, but it is less common. WSJT does it automatically for FT8 operation, to avoid problems caused by harmonics of the modulating audio signal.

Split operation in the FTDX10 is different from many transceivers where a pre-defined 5 kHz or similar offset is applied. Using Split in the FTDX10 means that the transceiver will transmit on the frequency indicated by the VFO-B display. That might be on a completely different band and mode, so you do have to be careful. Split operation is indicated by changing the VFO-B numbers to red and the red TX LED moves to VFO-B, above the CLAR RX button.

TIP: unlike the FTDX101, there is a TXW button that you can press to hear what is on the proposed VFO-B transmit frequency.

➢ **Tuning in Split mode**

Tuning in the split mode couldn't be easier! The main VFO knob tunes the VFO-A (receiver) frequency, and the VFO ring tunes the VFO-B (transmitter frequency). Pressing the SPLIT button automatically turns off the CS button.

Holding down the TXW button lets you listen to the frequency that the radio will transmit on. It is easier to see this on the spectrum scope if you have the markers turned on and use the CURSOR or FIX display setting.

➢ **Quick split input**

If you set QUICK SPLIT INPUT to OFF (default), and you press and hold the SPLIT button, the VFO-B mode is set the same as VFO-A and the frequency is set to the VFO-A frequency plus the QUICK SPLIT FREQ offset.

If you set QUICK SPLIT INPUT to ON, and you press and hold the SPLIT button, a dialogue box pops up for you to set the positive or negative split offset you want, in kHz.

TIP: I prefer the OFF setting, it is so easy to adjust the split (transmitter) frequency using the VFO tuning ring, that it seems cumbersome to have to type in the Split offset.

<FUNC> <OPERATION SETTING> <GENERAL> <QUICK SPLIT INPUT>

➢ **Setting the 'quick split offset'**

<FUNC> <OPERATION SETTING> <GENERAL> <QUICK SPLIT FREQ> sets the 'quick split' offset. The default offset is +5 kHz, but it is adjustable from -20 kHz to +20 kHz using the FUNC knob. If QUICK SPLIT INPUT is set to off and you press and hold the SPLIT button, VFO-B is moved by the amount you selected. If you press and hold the SPLIT button again, a further offset of the same amount is applied. For example, +5 kHz, +5 kHz, +5 kHz.

➢ **SPLIT button rules**

1. Pressing SPLIT only changes the transmitter to VFO-B frequency and turns the numbers red. It does <u>not</u> change VFO-B to be on the same band and mode as VFO-A.

2. If the transmitter was already on VFO-B, pressing SPLIT changes the transmitter to the VFO-A frequency and turns the VFO-A numbers red. It does not change the VFO-A band and mode.

3. Press and holding the split button when QUICK SPLIT INPUT is set to ON, opens a dialogue box for you to set the positive or negative split offset you want, in kHz. <5> <kHz>. It also corrects the VFO-B mode and band.

4. Press and holding the split button when QUICK SPLIT INPUT is set to OFF and Split is OFF, sets VFO-B to VFO-A mode and frequency plus or minus the pre-set split offset.

5. Press and holding the split button when QUICK SPLIT INPUT is set to OFF when Split is already turned ON, increments the split offset on VFO-B by the pre-set QUICK SPLIT FREQ offset.

➢ **A typical scenario**

For a typical scenario where I am attempting to contact a DX station using SSB on the 20m band. I recommend the following technique.

- Set QUICK SPLIT INPUT to OFF and QUICK SPLIT FREQ to +5 kHz using the instructions above. These are the default settings, so you only need to do this if you have changed them previously.

- If the markers are off, turn them on with <FUNC> <MARKER>.

- I recommend using the CURSOR or FIX display.

- Tune to the station that you want to call.

- Press and hold the SPLIT button to set VFO-B to the same band and mode as VFO-A with an offset of +5 kHz. *[This will not work if Split is already turned on]*.

- Press the A/B button to reverse the VFOs so that you can hear the frequency you have selected for transmitting. Use the VFO knob to find a suitable frequency then press the A/B button again.

- Make sure that you will be transmitting within the span of frequencies the DX station is listening to, for example, "5 kHz up," or "up 5 to 10 kHz."

- If you want to make a quick check of the transmit frequency you can press and hold the TXW button and use the main VFO knob to adjust the frequency that you will transmit on when using Split. You can also see the pile-up activity on the spectrum scope and waterfall display.

- Unlike the A/B button, the TXW button swaps the receiver to the other VFO but not the transmitter. It does not latch. You have to hold it down.

- The red marker shows your transmitter frequency. It will jump to the center of the display while you are transmitting. On the CENTER display, the center line shows the receiver frequency. On the FIX or CURSOR display, the green marker shows the receiver frequency

- Now you are ready to make your call.

The technique is similar when you are using the low bands except the pileup will be spread on frequencies below the DX station and you will use LSB. For CW operation, the technique is the same except the split will be 1-2 kHz rather than 5-10 kHz.

OPERATING DIGITAL MODES (FT8 ETC.)

Digital mode operation is easy once you have established the connection between the radio and the computer and set up the com port and audio settings.

FT8 has swept the world as it has allowed modest-sized stations the opportunity to work DX when band conditions have been marginal due to the sunspot cycle. As the new cycle ramps up over the next few years, I hope we will see many people return to digital 'conversation' modes such as PSK and RTTY.

I like to set up all my digital mode programs so that they all use the same settings on the radio. It makes changing between them so much easier. I use MixW for digital modes, WSJT-X for FT8, and MRP40 for CW. I can't go into the operation of these programs. That would be a whole other book! But I have included setup instructions for some of the popular programs in the 'Setting up PC digital mode software' chapter.

➢ **Soft Keys**

I usually use; ATT = OFF, IPO = AMP1, AGC = AUTO, R.FIL = 3 kHz. You could use the 500 Hz roofing filter for internal PSK or RTTY operation.

➢ **Span**

I usually use a 5 kHz to 10 kHz spectrum span when using Digital modes

➢ **Operating with external digital mode software**

Turn on your digital modes program. Select your operating frequency. Press MODE and select the mode. Usually DATA-U for external digital mode programs.

If you have set up the Preset mode, press MODE again and select Preset. The icon must be blue (not grey).

WSJT-X and some other programs have band dropdown menus or buttons which will automatically place the transceiver on the correct frequency.

If 'Rig' is selected, WSJT-X will use VFO-A for receiving and VFO-B for transmitting. Automatically shifting the transmitter frequency so that the modulating audio can be between 1500 to 2000 Hz. This stops harmonics of the audio signal creating false FT8 signals.

Make sure that the audio level into the digital modes program is sufficient, but not overloading the program. If necessary, use the program's soundcard settings or the Windows sound settings to set a suitable level. There is no audio output level on the transceiver.

Make sure that the transmit audio level into the radio is sufficient to modulate the radio to full power (or a lower power if you prefer). The ALC meter should be very zero or very low indicating that there is no compression of the digital mode signal. If necessary, use the program's soundcard settings or the Windows sound settings to set a suitable level. You can adjust the RPORT GAIN audio level on the transceiver, but I prefer to leave it set around 12. All of my digital mode software is set up using the individual program settings to avoid having to change settings on the radio all the time.

TIP: It is better to turn down the level on the digital mode software and/or the Windows sound settings than to leave the soundcard 'Playback' level at 50% or 100% and rely on the RPORT GAIN control to set the modulation level. That way you will avoid the embarrassing situation of generating audio harmonics in the sound card and transmitting your digital mode signal on two frequencies simultaneously.

The digital mode program should control the PTT via the RTS line on the Standard com port. Enjoy operating your favorite digital mode.

➢ **Operating with the internal decoders**

You don't need a connection with the PC to use the three modes that have internal decoders and message macro keyers. And there are no audio level settings to worry about. Press MODE and select the mode you wish to use, PSK, RTTY-L, or CW.

You do not need to use the Preset for the internal digital modes. Press and hold MODE to check that the Preset Soft Key icon is grey (not blue). If necessary, touch PRESET to turn it off.

On CW you can use ZIN and the tuning indicator on the filter function display to 'tune in' a signal and then apply the APF filter to remove most of the background noise. On RTTY use the mark and space tuning indicator lines on the filter function display to tune into an RTTY signal. On PSK tune the signal to align with the 'C' marker.

Now open the decoder screen using <FUNC> <DECODE>. It will overlay the spectrum and waterfall display. You can adjust the DEC LVL to minimize rubbish decodes but you have to be very, very, fast! The DEC LVL popup window only stays active for two seconds. To close the decoder, touch the DEC OFF Soft Key.

When you are ready to transmit one of the five messages. Press <FUNC> <MESSAGE>. Unfortunately, this will sit over the top of the Decode screen so you can't see the message that you are responding to. If you are planning on operating this way regularly you should definitely invest in an FH-2 keyboard. Or build your own using the schematic in this book. You can do a quick edit to add a callsign or other information by touching MEM then the message number.

Touching BACK, or any external Soft Key, or most buttons, will close the message keyer screen. Weirdly pressing FUNC resends the last sent message instead of exiting the screen. This is unnecessary since it is just as easy to touch the message Soft Key. I find it counter-intuitive and very annoying.

The RTTY and PSK keyers do not require you to have BK-IN turned on. In fact, the BK-IN button is disabled in the Data, PSK, and RTTY modes. See setup on page 16.

FM REPEATER OPERATION

Repeater operation is best achieved by tuning in a repeater, setting all the relevant offset and tone squelch requirements, then saving the channel into a memory slot. When you set a repeater offset or tone frequency in VFO mode, it will stay set, over the whole band, until you change it back. There is no auto-repeater mode on this radio.

Use the MODE button to select FM and tune the VFO to the repeater output frequency.

Press <FUNC> <RADIO SETTING> <MODE FM> and check that the following settings match the repeater that you want to access.

<RPT SHIFT (28 MHz)> 100 kHz default. The repeater offset for the 28 MHz (10m band) can be set in 10 kHz increments between 0 and 1000 kHz. But the default 100 kHz offset will suit the vast majority of repeaters.

<RPT SHIFT (50 MHz)> 1000 kHz default.

The FM offset for the 50 MHz (6m band) can be set in 10 kHz increments between 0 and 4000 kHz. The default 1000 kHz offset will suit the vast majority of repeaters. All New Zealand 6m repeaters use a -1 MHz offset, but some other countries use a 500 kHz offset.

RPT

There is a choice of, SIMP (simplex no offset), + (plus offset), and – (minus offset). With a plus offset the transmitter will transmit higher than the receive frequency. With a minus offset the transmitter will transmit lower than the receive frequency.

➢ **Setting the TONE**

Most amateur radio repeaters use a CTCSS (continuous tone coded squelch system) tone to prevent the repeater from being held on by noise or interfering signals. They usually transmit the same tone on the repeater output frequency so that you can enable tone-controlled squelch of your receiver.

Use <FUNC> <RADIO SETTING> <MODE FM> <ENC/DEC> to enable the CTCSS tone. There is a choice of, OFF, ENC, or TSQ.

- OFF no tone is transmitted.

- ENC a tone is transmitted to open the repeater squelch.

- TSQ a tone is transmitted to open the repeater squelch and the same tone must be received to open the squelch on the FTDX10.

The 'sub-audible' CTCSS tones are all between 67 Hz and 254.1 Hz. Which is below the normal 300 Hz – 3 kHz audio range of your receiver. This means that they can be used for signaling, but you won't hear them unless you reduce the LCUT FREQ of the audio low pass filter. The most often used tone is 67 Hz, but your local repeaters may use different tones.

Set the tone using, <FUNC> <RADIO SETTING> <MODE FM> <TONE FREQ> and turning the FUNC knob.

CTCSS Tones (Hz)							
67.0	69.3	71.9	74.4	77.0	79.7	82.5	85.4
88.5	91.5	94.8	97.4	100	103.5	107.2	110.9
114.8	118.8	123.0	127.3	131.8	136.5	141.3	146.2
151.4	156.7	159.8	162.2	165.5	167.9	171.3	173.8
177.3	179.9	183.5	186.2	189.9	192.8	196.5	199.5
203.5	206.5	210.7	218.1	225.7	229.1	233.6	241.8
250.3	254.1						

➢ **Operating through a repeater**

OK, so you have set the FM mode, tuned to the repeater output frequency, turned on the tone squelch, and set a plus or minus offset. You also ensured that the offset frequency is correct for the repeater and set the CTCSS tone frequency.

- With a plus offset your transmitter will transmit higher than the receive frequency because the repeater's input frequency is higher than its output frequency.

- With a minus offset your transmitter will transmit lower than the receive frequency because the repeater's input frequency is lower than its output frequency.

Now you can transmit and hopefully hear a 'kerchunk' when you click the microphone PTT. (Assuming the repeater has a 'tail.' Not all of them do).

TIP: All these settings are a pain to set up and they stay set, so they are still active when you tune down to a different part of the band. It is highly recommended that once you have verified that you can trigger the repeater, you save the current settings to a memory channel. Then deactivate the offset back to Simplex and set the tone squelch back to OFF.

➢ **Memory channels and FM**

There are no specific memory slots for FM channels. You can use one of the standard memory channels and name it with the repeater identification.

To store a channel: Tune to the repeater output frequency and set the offset and tone as above. Press M to bring up a list of memory channels. Use the FUNC knob or touch the memory slot you want to use.

Hold down M until you hear a double beep, and you will see the frequency pop into the selected memory slot.

Before you exit, there are some changes you can make.

- Touch NAME to enter the repeater's name. Remember to exit the keyboard with <ENT> to save the name.

- Touch MODE to use the FUNC control to change the mode. It should already be correct, so I don't know why you would want to do that.

- Touch SCAN MEMORY to include the channel in the scan mode or skip it. Use the FUNC knob to change the entry, (SCAN or SKIP)

- Touch DISPLAY TYPE to display the stored name instead of the frequency in the VFO frequency display. I rather like this one. Use the FUNC knob to change the entry, (FREQ or NAME).

- Exit using BACK or by pressing FUNC.

➢ **FM and FM-N modes**

Use the FM mode for FM operation rather than the FM-N mode, unless you are using the radio with a transverter for satellite operation or on frequencies where the narrow band FM-N mode may be required.

The I.F. bandwidth is fixed at 16 kHz on the FM and DATA-FM modes and 9 kHz on the FM-N and D-FM-N modes. The transmitter deviation is ± 5kHz on the FM and DATA-FM modes and ±2.5 kHz on the FM-N and D-FM-N modes.

You cannot adjust the roofing filter or the I.F. bandwidth while the radio is in FM mode. The roofing filters are switched out leaving a 12 kHz RF passband.

COMPUTER MOUSE OPERATION

Connecting a computer mouse is easy. Just plug a USB mouse into one of the front panel USB ports and you are ready to go. I have been told that the Logitech M310 and M305 wireless models will work, and probably some other mice. The right mouse button has no function and sadly the mouse wheel does nothing either. It could have been used to scroll the frequency... but no.

So, what can you do with the mouse? These items are not in the manual.

- The left mouse click has the same effect as touching the screen with your finger. You can change the meters, the filter function display, the waterfall to spectrum ratio, the frequencies, and use the Soft Keys.

- If you press FUNC, you can use the mouse to select any of the menu functions. USB Keyboard operation

TIP: if you use the horrible external display you will need a mouse to make selections.

You can connect a USB keyboard to the radio and use it to fill in the TEXT messages. However, I cannot see much point in plugging an external keyboard when there is an onscreen keyboard provided for this purpose. Unfortunately, you can't use an external keyboard in conjunction with the decoders to send text directly from the keyboard.

Another odd feature is that if I plug in my Logitech wireless keyboard and mouse combo. The keyboard works, but the mouse doesn't.

SCAN MODE

You can initiate a scan by press and holding either of the microphone UP or DWN buttons. Press again to stop the scan or touch the spectrum display or click the microphone PTT. This function can be turned off with <FUNC> <OPERATION SETTING> <GENERAL> <MIC SCAN> (ON or OFF).

A single press of either of the microphone UP or DWN buttons steps the VFO by 10 Hz (or 5 Hz), the same as tuning the VFO. In FM it will step by the nominated step size, in my case 25 kHz.

Starting a scan while in the V/M memory channel mode scans through the saved memory channels. Channels can be 'skipped' (excluded from the scan), by changing the setting in the memory channel list. Press M and use the FUNC knob, to find the channel, then change the SCAN MEMORY setting from SCAN to SKIP. A single press of either of the microphone UP or DWN buttons steps the VFO to the next memory channel.

TIP: You can change the scan direction by rotating the VFO tuning knob about a quarter of a turn during a VFO scan. Anti-clockwise tunes down, clockwise tunes up. This is useful if you scan over a signal and want to go back to it.

Set the squelch control so that the receiver audio is just muted. In the SSB and CW modes, the scan rate will slow down if a signal opens the squelch. In the other modes, the scan will stop if a signal opens the squelch. You can set what happens when the scan stops. TIME starts the scan again after 5 seconds. PAUSE halts the scan until the signal drops out and then the radio resumes scanning. <FUNC> <OPERATION SETTING> <GENERAL> <MIC SCAN RESUME> (PAUSE or TIME).

PROGRAMMABLE MEMORY SCAN (PMS)

The radio supports up to nine pre-programmed scan ranges, using nine pairs of 'special-purpose memory pairs' known as L/U pairs.

➢ **Setting up an L/U pair**

Tune to the lower end of the frequency range that you want to save.

Select M to activate the MEMORY CH LIST. Use the FUNC knob to navigate all the way down to the bottom of the 99 memory channels until you reach the M-P1L and M-P1U memory pair.

Touch the M-P1L slot, or whichever L/U slot you want to program. Make sure that you have the right mode selected, then press and hold the M button to store the current VFO frequency into the slot. You can name the range if you want to.

Now tune the VFO up to the high end of the frequency range that you want to save.

Touch the M-P1U slot, or whichever L/U slot corresponds to the L/U pair you have chosen. Press and hold the M button to store the current VFO frequency into the slot. Exit using FUNC if you want to load the frequency into the VFO or BACK if you don't.

➢ **Starting a PMS scan**

Select M to activate the MEMORY CH LIST. Navigate down to the L/U channels using the FUNC knob.

Select the M-P_L channel that you want, then press the FUNC knob to set the radio into the memory channel mode. This bit is unusual and very important. Turn the VFO knob ¼ turn to set the radio to PMS. It will be indicated above the MODE indicator. Then press and hold the UP button on the microphone to start the scan.

Pressing the DOWN button on the mic during the scan will make the radio scan down.

Pressing the UP button on the mic during the scan will make the radio scan up. You can also change the direction of the scan by turning the VFO knob a little in the direction that you want the scan to go.

When a signal opens the squelch, the scan will slow on SSB or CW and stop on the other modes.

Touch the spectrum or waterfall display or click the microphone PTT to stop the scan. To escape the PMS mode press V/M to get back to M-P1L, then press V/M again to get back to the VFO mode.

LIMITING THE TUNING RANGE

The PMS mode described above limits the range that you can tune the VFO. For example, if you set a PMS L/U pair for 3.750 MHz to 3.900 MHz, and activate PMS (without starting a scan), you will only be able to tune over that part of the 80m band. This could be handy during a contest as it would stop you from tuning outside of the band of interest. But it only lasts until you change bands. When you return the 80m band the radio is no longer in PMS mode. It reverts to the MT memory tune mode, which is not constrained to the PMS limits.

TIP: If the radio has switched to MT mode, you can press V/M to return to M-P1L. Turn the VFO clockwise and the radio will go back to PMS. Or press V/M again to get to VFO mode.

➤ **Using PMS to limit the tuning range**

Select M to activate the MEMORY CH LIST. Navigate down to the L/U channels using the MULI knob.

Select the M-P1L, or another PMS scan edge and press the FUNC knob to set the radio into memory channel mode. Then turn the VFO knob clockwise to set the radio to PMS. It will be indicated above the orange MODE indicator. Now you can use the VFO as normal, and the radio will not let you tune outside of the PMS range.

To escape the PMS mode press V/M to get back to M-P1L, press V/M to get back to VFO. Or change bands.

USING THE ANTENNA TUNER

The internal tuner is able to match unbalanced loads between 16.7 and 150 ohms. This is equivalent to a SWR of less than 3:1. Or 25 to 100 ohms on the 6m band which represents an SWR less than 2:1. These tuning ranges are typical for antenna tuners installed inside HF transceivers and will let you match out to the band edges on 'resonant' antennas such as a quarter-wave vertical, a half-wave dipole, or a Yagi antenna. The tuner will struggle to tune highly reactive loads. The tuner does not operate on the 70 MHz (4m) band. Turning it on can result in high SWR when you transmit.

You should not use the internal tuner and a non-Yaesu external tuner at the same time. An external antenna tuner or a linear amplifier followed by an external tuner should present the radio with an acceptable termination. Leave the internal antenna tuner turned off.

Pressing TUNE will turn the antenna tuner on. A red 'TUNE' icon will appear at the top of the spectrum display (under the four Soft Keys) to indicate that the tuner is active. The tuner will choose settings that were saved after an antenna tuning operation was performed on a frequency within the 10 kHz window. If no settings have been saved for the current frequency, the tuner will choose settings that were saved after a tune operation was performed on a frequency close to the current frequency.

Note that the tuner will not automatically tune if you transmit into a high SWR. You must initiate the tuning operation by press and holding the TUNE button.

Press and hold TUNE to activate a tuning sequence. This will radiate some power into the antenna, so please make sure nobody is using the frequency. The tuning sequence takes a few seconds. The radio will always perform a full tune, ignoring any previously saved information. If you have the SWR meter active you will see the SWR come down as the tuning operation nears completion.

TIP: It is not necessary to do a full tune every time you transmit on a different frequency. The tuner will automatically return to the settings it used the last time you transmitted near the selected frequency. It stores tuning information for each 10 kHz window. Check the SWR while you are transmitting. If it is high start a TUNE operation.

➢ **ATU memories**

The antenna tuner will store the tuning settings after a tuning operation has been completed, provided that the final SWR is less than 2:1.

On the HF bands, if the resulting SWR after the tuner has attempted to match to the antenna is higher than 2:1 but less than 3:1 the tuner will be able to match the load. But the settings will not be stored.

If the SWR is greater than 3:1 the tuner may not be able to find a match.

If you change to a frequency that is outside of the 10 kHz band segment, the tuner will revert to any previously saved settings for the new band segment, or to previously saved settings for the closest band segment.

➢ **Rules for the antenna tuner**

1. Check the SWR meter while transmitting. If the SWR is high, you should stop transmitting and start a 'tune' operation by press and holding the TUNE button.

2. Pressing the TUNE button will turn the tuner on. It will set the tuner to capacitance and inductance settings that previously resulted in a match. If the antenna is the same and you use a frequency that the antenna tuner has tuned before, you should have a well-matched system. Check the SWR meter while transmitting to make sure that the meter hardly moves.

3. The antenna tuner will not automatically initiate a 'tune' operation if the SWR is still high. You have to manually start a 'tune' by press and holding the TUNE button.

➢ **External Yaesu antenna tuner or ATAS antenna**

If you are using a Yaesu FC-40 antenna tuner, or an ATAS 120A active tuned antenna, you need to change a menu setting in the radio. Set <FUNC> <OPERATION SETTING> <GENERAL> <TUNER SELECT> to (EXT or ATAS).

The Yaesu manual has information about connecting a Yaesu FC-40 remote mounted external antenna tuner. Connecting an FC-40 or ATAS 120A automatically disables the FTDX10 internal tuner. Note the requirement for installing a ferrite core isolator at the transceiver end of the control cable. See page 108 of the Yaesu Operating Manual. © Yaesu.

USING A LINEAR AMPLIFIER

➢ **Yaesu VL-1000 linear amplifier**

If you are connecting a Yaesu VL-1000 linear amplifier, the PTT switching, automatic band switching, and ALC control are all managed over a CT-118 cable. It connects the 10 pin 'Linear' connector on the FTDX10 to the 'Band-Data 2' and 'ALC 2' connectors on the amplifier. The connections are covered in the Yaesu Manual. You should set up the ALC as covered below.

➢ **Other linear amplifiers**

Rather unusually the FTDX10 does not have RCA jacks for TX GND (PTT) and EXT ALC. If you are connecting a non-Yaesu linear amplifier, you can use the TX GND and EXT ALC control signals available on the 10 pin 'Linear' connector. The Yaesu band data is also available. See 'Rear Panel Connectors,' on page 169.

➢ **Linear amplifier ALC**

The ALC (automatic level control) voltage sent from the linear amplifier to the transceiver automatically reduces the transmitted power so that the linear amplifier is never overdriven, which could cause audio compression or clipping, a protection trip, or possible failure of the linear amplifier. The ALC should be configured so that it is not operating unless the transceiver is accidentally left at full power when driving the amplifier. Do not use it as a method of controlling the amplifier power.

It should only be used as a failsafe in the event of a power setting mistake. The FTDX10 ALC line accepts a voltage between 0 and -4 Volts. Check that your amplifier has a compatible output level.

Note: Linear amplifier ALC is completely different from the microphone ALC measured by the ALC meter. Microphone ALC is used to ensure that the modulator is not overloaded by your speech when you are using SSB.

➤ **Adjusting the linear amplifier ALC level**

The common practice of running your transmitter at full power and using the linear amp ALC to ensure that the linear amplifier is not overdriven is a bad idea.

The linear amplifier ALC is supposed to be used as a protection method. It is too slow to react to sudden changes in input level or help much with transmitter overshoot.

This adjustment requires you to transmit at the full power of the linear amplifier. If you have a high-power load you can test into that. Otherwise, use your normal antenna and conduct the test at a time and on a frequency that won't interfere with other users.

1. Set the transmitter power well below the level required to drive the linear amplifier to full power. <FUNC> <RF POWER>.

2. Set the radio to a mode that will transmit at the full power set above, such as CW, FM, or PSK. If you are using the preferred CW mode, turn off the Keyer.

3. Set the ALC control on the linear amplifier fully off, (probably fully clockwise). It may be a potentiometer on the rear panel of the amplifier, a menu command on the amplifier, or a software control.

4. Transmit into the linear amplifier and slowly increase the RF power until the linear amplifier reaches full rated power. Stop transmitting for a while to give the amplifier a rest.

5. While transmitting, change the ALC control until the output power of the transmitter and linear amplifier just begins to reduce. Then back off the control slightly until the linear amp is back to full power and the ALC is not having any effect. At that point, the ALC is not operating during normal transmission but will respond to protect the linear amplifier if you accidentally leave the transmitter at full power.

TIP: Many older linear amplifiers need 100 watts of drive to reach full output power. The same ALC setting procedure can be applied, but it should be impossible to overdrive the amplifier anyway. I think that it is users of this older type of amplifier that started the trend of running full transmitter power and using the ALC to reduce the power to avoid overdriving the amplifier.

RECORDING AND PLAYBACK

An SD card must be in the radio. Touch <FUNC> <RECORD> to start recording the receiver audio. An S.REC icon flashes below the MODE indicator. The RECORD Soft Key changes to STOP. Recording continues until you touch <FUNC> <STOP>.

Touch <FUNC> <PLAY> to play back or delete your recorded files. The command opens a PLAY LIST with standard 'tape recorder' controls. Select a file and touch the play icon to play a file. The table includes the filename, which usually contains the date and time of the recording. The center column shows the length of the recording, and the final column displays the mode. An indicator at the top shows the elapsed and remaining time of the recording as it is being played back.

|◀◀   moves to the previous saved file

▶▶| moves to the next saved file

▶   play

◀◀   rewind

▶▶   fast forward.

RX LEVEL changes the playback volume. 20% seems good.

DELETE – yep you guessed right!

The files are stored as .wav files in the PlayList subdirectory on the SD card.

Memory groups and channels

The FTDX10 has 99 normal memory slots which can be arranged into five groups. There is a 6th group allocated to the nine L/U scan ranges, see PMS on page 92. The U.S. (International) version of the radio also stores ten 5 MHz frequencies which cannot be erased. The U.K. model stores seven 5 MHz frequencies. Memory channel one (M-01) cannot be erased, but it can be overwritten with a new frequency and mode. However, if the radio is reset, M-01 will return to the default setting of 7.000 MHz LSB.

➢ **Saving to a memory channel**

Saving a frequency to a memory channel is very easy. Set the radio to the mode and the frequency you want to save. For FM repeaters, tune to the repeater output frequency and set the repeater offset and CTCSS tone.

Press M to bring up a list of memory channels. If you want to save the frequency within a group, check out the section on groups below.

Use the FUNC knob to find a suitable memory position. Or touch the wanted memory slot. Hold down the M button until you hear a double beep and see the frequency pop into the memory slot.

Before you exit, there are some changes you can make.

- Touch NAME to enter the repeater, net, or channel name. Remember to exit the keyboard with <ENT>.

- Touch MODE and use the FUNC control to change the mode. I don't know why you would want to do that, but you can.

- Touch SCAN MEMORY to include the channel in the scan mode or skip it. Use the FUNC control to change the entry, (SCAN or SKIP).

- Touch DISPLAY TYPE to display the stored name instead of the frequency in the VFO frequency display, (FREQ or NAME).

Press the M button again to return to the normal display. Or press the FUNC knob to switch to the memory channel mode and load the selected memory channel into the VFO.

Note that pressing V/M will swap the radio between the VFO mode and the memory channel mode, but it will not close the MEMORY CH LIST.

➢ **Recalling a saved frequency**

Recalling a saved frequency from a memory channel is easy as well.

Press M to bring up the list of memory channels.

Use the FUNC knob to find the saved memory channel you want to select. Or touch the relevant memory slot. When the channel you want is highlighted, press the FUNC knob to switch to the memory channel mode and load the selected memory channel into the VFO.

> **Turning on and using the memory groups**

Use <FUNC> <OPERATION SETTING> <GENERAL> <MEM GROUP> (ON or OFF) to enable or disable the memory group function.

You can't just allocate a memory slot to a particular group. To save a frequency to a group you have to save it into a memory slot that is within the group.

Channels M-01 to M-19 are in Group 1

Channels M-20 to M-39 are in Group 2

Channels M-40 to M-59 are in Group 3

Channels M-60 to M-79 are in Group 4

Channels M-80 to M-99 are in Group 5

PMS channels P1 L/U to P9 L/U are in Group 6

You can force the memory mode to only display memory channels within a certain group. But the process is so complicated I doubt that many people will bother. It is probably better just to arrange your most used channels near the top of the memory stack.

Ensure that 'memory groups' are turned on, as mentioned above.

Press V/M to change to the memory channel mode.

Press <FUNC> <M-GROUP> and then use the FUNC knob to select a group.

TIP: this will only show groups that have frequencies saved in them, including the PMS scan group and the 5 MHz channels.

Click the microphone UP and DWN buttons to step through the saved memory sots in the group. If you press and hold either of the microphone UP and DWN buttons the radio will scan the memory channels in the group.

> **MT memory tune mode**

If you select a memory channel and then turn the VFO knob, the radio will tune off the saved channel frequency in the same way as when you are using the VFO mode. This is indicated with MT in place of the M-xx channel number above the mode display.

Pressing the V/M button returns the radio to the saved memory channel frequency.

Pressing the V/M button again returns the radio to the VFO mode.

➤ **Stepping through saved memory channels**

Press V/M to select the memory channel mode. Then press and hold the STEP/MCH button. This lets you step through the memory channels by tuning the VFO tuning ring. If 'memory groups' are turned on, you can only step through channels in the currently selected group. The MCH mode is indicated with a flashing orange LED on the STEP/MCH button. *In the VFO mode, press and holding the STEP/MCH button changes the VFO tuning ring tuning rate to 1 MHz.*

Or press V/M to select the memory channel mode and click the microphone UP and DWN buttons to step through the saved memory sots. If you press and hold either of the microphone UP and DWN buttons the radio will scan the memory channels.

➤ **Memory channel scan**

You can start a memory channel scan by press and holding the microphone UP or DWN button, while in the memory channel mode. Channels can be 'skipped' (excluded from the scan), by changing the 'Scan Memory' setting in the memory channel list. Press the M button and use the FUNC knob, to find the channel, then change the SCAN MEMORY setting from SCAN to SKIP. If a memory group has been selected only 'non skipped' channels in that group will be scanned. If MEM-GROUP is turned off, all 'non skipped' channels will be scanned.

➤ **Backup the memory list to the SD card**

Firmware updates or resetting the radio may delete your saved memory channels, so it is a good idea to save a backup onto the SD card. Make sure that there is an SD card in the SD card slot. Then select, <FUNC> <EXTENSION SETTING> <SD CARD> <Mem List Save DONE> <NEW> <ENT>. You can change the filename if you want to, but I just touch ENT to select the default date format as the filename. The radio will format the data file and save it to the SD card. Touch the 'FILE SAVED' icon to return to the previous menu.

You might as well save a backup of the menu settings at the same time. Touch <Menu Save DONE> <NEW> <ENT>. Then press BACK, BACK to exit.

➤ **Recalling the memory list from the SD card**

If a radio reset, or a firmware update, wipes out your saved frequencies, you can recover the memory list from your saved backup on the SD card. Make sure that the SD card is in the SD card slot. Then select, <FUNC> <EXTENSION SETTING> <SD CARD> <Mem List Load DONE>. You will be presented with a list of previously saved files. Unless there has been a data corruption problem, you would normally select the newest file, which will be at the bottom of the list. Touch the file name and answer 'OK' to overwrite the memory slots in the radio with the ones stored in the selected file. Touch the 'FILE LOADED' icon and the radio will re-boot to load the memory channels.

LED indicators

The LEDs across the top of the control area indicate important status information.

			RX	TX	RX	TX	RX	TX	FINE	STEP
Receiver squelch	Transmit		Memory mode		VFO-A		VFO-B		Tuning Step	Tuning Rate

BUSY	Green	The radio is receiving a signal, or the receiver squelch is open. Only active when the RF/SQL knob is set to squelch.
TX	Red	The radio is transmitting
MEM	Green	Memory channel mode is selected for receiving
MEM	Red	Memory channel mode is selected for transmitting
VFO-A	Green	VFO-A is selected for receiving (Normal)
VFO-A	Red	VFO-A is selected for transmitting (Normal)
VFO-B	Green	VFO-B is selected for receiving
VFO-B	Red	VFO-B is selected for transmitting (Split)
FINE	White	The VFO and VFO tuning ring are in 'fine tuning' mode
STEP	White	The VFO tuning ring is in step tuning mode

ADJUSTING THE LED BRIGHTNESS

The LED Dimmer adjustment sets the brightness of the LEDs, from 0 to 20. This is a nice feature. You may want to dim the LEDs if you are working at night, especially if it's a contest and your eyes are tired. I set my radio for 7, a little lower than the default level of 10.

<FUNC> <DISPLAY SETTING> <DISPLAY> <LED DIMMER>

ADJUSTING THE SCREEN BRIGHTNESS & CONTRAST

Pressing <FUNC> <DIMMER> and using the FUNC knob sets the brightness of the touchscreen between 0 and 20. (Default 10).

Pressing <FUNC> <CONTRAST> and using the FUNC knob sets the contrast of the touchscreen between 0 and 20. (Default 10).

Touchscreen display functions

This section deals with the interactive items that are displayed on the touchscreen. It describes what happens when you touch the meter displays, the frequency readout, and the spectrum scope. The next section covers the Soft Key controls.

METER

While the radio is receiving, the meter always shows the received signal in S points.

Touch the meter to select the transmitter meter options. It can display; PO (power out), COMP (speech compression), ALC (automatic level control of the modulating audio), V_{DD} (power amp FET drain voltage), I_D (power amp FET drain current), or SWR (antenna standing wave ratio).

Figure 28: Meter display

FILTER FUNCTION DISPLAY

The 'filter function display' shows the received spectrum within the receiver passband. The white line outlines the shape of the filter response. The bandwidth of the currently selected R.FIL roofing filter is indicated by the blue line at the bottom of the spectrum display and a white background shading. You can see the shaded area more easily if you adjust the I.F. width or shift control quite a lot.

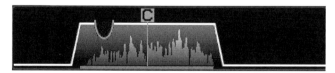

Figure 29: Filter function display

Touch and hold the filter function display to toggle the spectrum display inside the filter shape off or on.

An orange dip or peak indicates that the CONTOUR filter is turned on and the frequency that it is set to.

The display indicates the SSB passband on SSB. The 'P' marker on CW shows the correct tuning point for CW signals. A second vertical orange line indicates that the APF filter is turned on and the frequency that it is set to.

On the internal PSK mode or the DATA-U mode, the 'C' marker indicates the center frequency of the passband.

'M' and 'S' markers indicate the Mark and Space frequencies for the internal RTTY mode.

THE FREQUENCY DISPLAY

The VFO frequency is displayed in large white numerals. The alternate VFO is displayed in smaller numerals below the main VFO.

The current mode is displayed to the left with a white indicator above. The indicator reads VFO-A or VFO-B in the VFO mode, M-xx in memory channel mode, or MT in the memory tune mode. In the PMS (programmable memory scan) mode, it indicates the PMS channel number. As soon as you move the VFO tuning knob, this changes to PMS. This will limit the VFO tuning to the PMS memory range and a scan will only cover the same range.

If the radio is in memory channel mode and 'Display Type' has been set to display the name instead of the frequency, the channel name will be displayed instead of the frequency digits.

The frequency display has several touchscreen functions.

➢ **Hz digits**

Touch the three Hz digits to enter a frequency using the onscreen keyboard. Touch ENT to load the frequency into the VFO. You only need to enter digits to the resolution you want. For example, <141> <ENT> will set the VFO to 14.100 MHz.

TIP: If the entered frequency is above 10 MHz there is no need to touch the decimal point. In fact, the decimal point will not respond if a double-digit number has already been entered.

➢ **kHz digits**

Touch the three kHz digits to tune the VFO in 1 kHz steps. The display will blink and then drop back to the normal tuning rate a few seconds after you stop tuning the VFO.

➢ **MHz digits**

Touch the MHz digit(s) to tune the VFO in 1 MHz steps. The display will blink and then drop back to the normal tuning rate a few seconds after you stop tuning the VFO.

➢ **Tuning rates**

Normally for SSB or CW, with CS turned off, the VFO knob tunes the radio in 10 Hz steps, and the VFO ring tunes in 100 Hz steps. If CS is turned on, the VFO tuning ring will adjust the assigned function instead. Usually spectrum scope level.

The FINE tuning button decreases the VFO tuning rate by a factor of 10, to 1 Hz steps for SSB and the Data-U mode. The VFO tuning rate stays the same. 100 Hz for SSB. Indicated by 'FINE' on the LED strip and an orange LED on the FINE button.

The VFO tuning rate can be changed to 5 Hz per step for SSB/CW or RTTY/PSK.

<FUNC> <OPERATION SETTING> <TUNING> <SSB/CW DIAL STEP>

<FUNC> <OPERATION SETTING> <TUNING> <RTTY/PSK DIAL STEP>

The normal tuning step for AM or FM is 100 Hz. The VFO tuning ring tunes in 1 kHz steps, and the FINE tuning rate tunes in 10 Hz steps.

➢ **STEP DIAL rates**

If you have pressed the STEP/MCH button The VFO tuning ring tunes in 10 kHz steps. However, you can choose 1 kHz, 2.5 kHz, 5 kHz, or 10 kHz steps.

<FUNC> <OPERATION SETTING> <TUNING> <CH STEP> changes the step dial rate for all modes except AM and FM, to 1, 2.5, 5 kHz or 10 kHz. I have set this to 1 kHz per step as most SSB stations will space across the bands at 1 kHz intervals.

<FUNC> <OPERATION SETTING> <TUNING> <AM CH STEP> changes the step dial rate for AM, (2.5, 5, 9, 10, 12.5, or 25 kHz)

<FUNC> <OPERATION SETTING> <TUNING> <FM CH STEP> changes the step dial rate for FM, (5, 6.25, 10, 12.5, 20, or 25 kHz). I set mine to 25 kHz.

➢ **Optical encoder rate VFO**

You can also change the rate at which the VFO knob works by changing the number of steps per revolution. If you select a lower number of steps per revolution, the VFO step size remains the same, but you have to rotate the knob more degrees to make the numbers change. If you select a higher number of steps per revolution, the VFO step size remains the same, but you don't have to rotate the knob as much to make the numbers change. It effectively changes the sensitivity of the control.

<FUNC> <OPERATION SETTING> <TUNING> <MAIN STEPS PER REV> (250, 500, or 1000 steps per revolution).

➢ **Optical encoder rate MPVD "VFO tuning ring"**

To change the encoder rate for the MPVD (Multi-Purpose VFO Outer Dial), otherwise known as the "VFO tuning ring," select,

<FUNC> <OPERATION SETTING> <TUNING> <MPVD STEPS PER REV> (250 or 500 steps per revolution).

SPECTRUM AND WATERFALL DISPLAY

Touching the screen within the spectrum scope area moves the VFO to the frequency you touch. It provides a fast way to navigate to a wanted signal or frequency. You will still have to fine-tune using the VFO knob.

Touching the 2D waterfall display toggles through three options that set the ratio of the waterfall area to the spectrum scope area. Roughly 2/3 spectrum, 50/50, or 1/3 spectrum and 2/3 waterfall.

You can use a USB computer mouse to change the same things as touching the screen with your finger or a stylus.

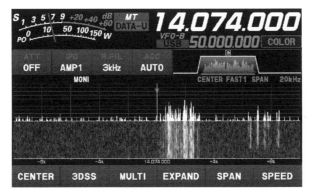

Figure 30: 2D spectrum and waterfall display

Figure 31: 3DSS spectrum

Display Soft Keys

This chapter discusses the operation of the Soft Keys. What they do and how to use them. There are six Soft Keys at the bottom of the display and four in the middle, above the spectrum and waterfall display.

THE LOWER SOFT KEYS

The six lower Soft Keys are used to set various display functions.

CENTER	3DSS	MULTI	EXPAND	SPAN	SPEED

Figure 32: The six lower Soft Keys

➢ **CENTER, CURSOR, and FIX**

The first Soft Key is used to control the spectrum scope and waterfall display. It toggles between the CENTER, CURSOR, and FIXED display modes.

The **CENTER** option makes the display act as a band-scope, like the displays fitted to superheterodyne architecture DSP radios. The receiver frequency is displayed in the center with a range of frequencies above and below.

NOTE: If you are using the normal 'carrier point' display, an SSB signal will be displayed to the right of center for upper sideband, or to the left of center for lower sideband. The displayed width depends on the selected roofing filter and the span of the display. If you selected 'Filter,' the spectrum and waterfall display will lie across the centerline. If markers are enabled a marker will indicate the carrier point. To toggle the markers on or off, select <FUNC> <MARKER>.

To change the scope center point, select <FUNC> <DISPLAY SETTING> <SCOPE> <SCOPE CTR> and select (CARRIER or FILTER). The FILTER option is horrible. Don't use it.

The **CURSOR** option behaves more like the spectrum and waterfall on an SDR receiver. Remember that the spectrum display is created by the direct sampling SDR. The point where the radio is receiving can be anywhere across the display.

Other than the received signals there is no indication of the receiver bandwidth on either the spectrum or waterfall display. This is a bit surprising as the FTDX101and the majority of other SDR radios do have this feature.

If the markers are turned on, a marker line will be shown on the spectrum display at the carrier point frequency. <FUNC> <MARKER> <ON>. The red marker indicates the transmitter frequency, and the green marker indicates the receiver frequency. In simplex mode, the red marker overlays the green marker.

You can see both markers if you select Split operation or use the clarifier. I strongly recommend leaving the markers turned on all the time.

If you tune above or below the range of frequencies in the currently displayed span, the markers will sit at the edge of the display and the entire spectrum and waterfall or 3DSS display will scroll as you tune. Touching the spectrum display will move the VFO to a frequency close to the position you touched.

The **FIX** display works in a similar way to the CURSOR display, except the behavior when you tune off the edge is different. When you touch FIX there is a good chance that the receiver frequency will not be shown on the display at all. Touching the spectrum or waterfall will move the VFO so that it is within the fixed display band.

FIX is the only lower Soft Key that has a touch and hold function. Touch and hold it to open a popup window that allows you to set the frequency that will be at the extreme left side of the spectrum display. Enter the frequency and then touch ENT to save your data. For example, <141> <ENT> will set the left side of the spectrum scope to 14.100 MHz. The frequency at the right side of the display is automatically adjusted according to the scope bandwidth set by the SPAN setting.

Unlike the CURSOR display which scrolls if you tune below or above the current scan. The FIX display stays, well… "fixed." If you have markers turned on, small green and red indicators at the right or left indicate which direction the VFO frequency is, compared to the displayed span. Touch the spectrum scope to move the VFO frequency back inside the fixed panadapter bandwidth.

The problem with the FIX display

The implementation of the FIX display is a major disappointment. I hope that Yaesu will look at the way that this mode is implemented on virtually every other SDR receiver and issue a firmware update.

You can only set a start frequency for the left side of the spectrum display. The frequency at the right side of the display is set according to the currently displayed Span setting. At a minimum, you should be able to set both the high and low, frequency limits for the fixed scope on each band.

As a comparison, the recent Icom transceivers allow you to set four fixed display regions for each of twelve frequency ranges. I have one set up to show the 20m band SSB segment from 14.100 to 14.350 MHz, the second shows the CW band segment from 14.000 to 14.060 MHz and the third displays the digital mode part of the band from 14.060 to 14.100 MHz. The only constraint on the Icom radio is you can't display a band segment that is wider than the 1 MHz maximum display bandwidth.

> ➤ **3DSS**

The 3DSS (three dimensions signal stream) is a spectrum display that includes signal history, similar to the information on a waterfall display. It is represented as a "3D" image tapering back to give an illusion of depth. Yaesu is very excited about this way of representing received signals, but I prefer the traditional spectrum and waterfall display. The COLOR setting applies to both the 2D normal spectrum and waterfall display and the 3DSS display. It has the same sensitivity as the 2D display.

The 3DSS Soft Key changes to blue when the 3DSS screen is active.

Adjusting the 3DSS display level

You will probably have to reduce the spectrum scope LEVEL to get a good picture. Adjust the level for the 3DSS display until the background noise is about a 50% mix of black and colored speckles. That way you maximize the dynamic range of the display. Signals should show as rows of spikes receding into the distance. The super-bright images on the Yaesu advertisements and many websites show the 3DSS display level set much too high.

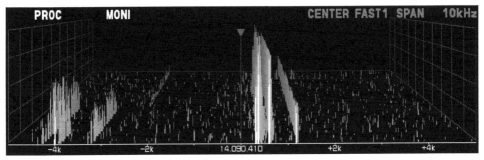

Figure 33: Adjusting the 3DSS display

TIP: If you adjust the 3DSS level but you want the background less bright, change the <FUNC> <DISPLAY SETTING> <SCOPE> <3DSS DISP SENSITIVITY> setting to NORMAL. You can also adjust <FUNC> <PEAK>.

2D and 3DSS sensitivity

There are 'HI' and 'NORMAL' sensitivity settings for the 3DSS and the 2D waterfall and spectrum display. Try both settings and see which you prefer.

<FUNC> <DISPLAY SETTING> <SCOPE> <3DSS DISP SENSITIVITY>

<FUNC> <DISPLAY SETTING> <SCOPE> <2D DISP SENSITIVITY>

➢ **MULTI**

The MULTI Soft Key turns on the audio oscilloscope and the audio spectrum display. When the radio is receiving you can see signals within the audio pass band. When transmitting you can see the quality of the transmitter audio.

TIP: If you are using one of the voice modes, you can listen to the transmitted audio using headphones at the same time as looking at the transmitted signal on the MULTI displays. Select <FUNC> <MONITOR> and turn the monitor level up with the FUNC knob.

The MULTI Soft Key changes to blue when the MULTI displays are active and the spectrum and waterfall or 3DSS display will shrink to fit.

You can change the measurement level and the time-base of the oscilloscope by touching the blue text at the top of the oscilloscope display. For speech, you might select 10 ms/division and 0 dB of attenuation. For CW try 100 ms or 300 ms/division and 10 dB of attenuation.

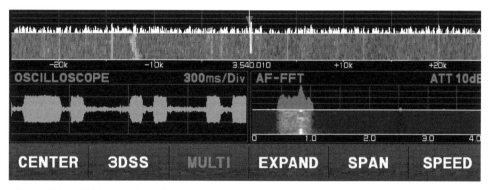

Figure 34: a CW signal on the MULTI display

You can change the sensitivity of the audio spectrum display by touching the blue text at the top of the audio spectrum display and selecting 0, 10, or 20 dB of attenuation. I found 10 dB to be a good option. The span is preset to 4 kHz.

➢ **EXPAND**

The EXPAND Soft Key makes the spectrum scope larger, at the expense of losing the filter function display and the upper Soft Keys.

➢ **SPAN**

The SPAN Soft Key sets the bandwidth of the spectrum and waterfall or 3DSS display. The maximum bandwidth is 1 MHz, but in most cases, you don't want to see that much of the band. You can set the displayed range to; 1 kHz, 2 kHz, 5 kHz, 10 kHz, 20 kHz, 50 kHz, 100 kHz, 200 kHz, 500 kHz, or 1000 kHz.

The best settings are a balance between the bandwidth of the signals that you are receiving and the spread of those signals across the band. CW signals are very narrow, and the CW segment of the band is usually 70 kHz or less, so you could use a 50 kHz or 100 kHz span. The PSK segment is smaller, but the signals are wider, so a 20 kHz or 50 kHz span is appropriate. I usually use 50 kHz for SSB, but if I am working a contest, I will increase the span to 200 kHz or 500 kHz so that I can see the contest stations across the band.

FT8 is crammed into a few kHz, and you do not have to tune the radio, because you use the digital mode software to select the stations that you want to receive. If you wanted to display the FT8 activity, you could set a span of 5 kHz or 10 kHz.

TIP: The FIX display is ideal for contesting using the "Search and Pounce" technique. Set the start frequency to the bottom of the SSB band segment, or CW segment for a CW contest, and set the span wide enough to include the whole band segment. For example, for an SSB contest on 20m, I might set the FIX to start at 14.150 MHz with a 200 kHz span.

> **SPEED**

The SPEED Soft Key sets the scrolling speed of the 2D waterfall and the 3DSS spectrum display. It does not affect the 2D spectrum display. I like to use FAST1.

THE UPPER SOFT KEYS

The radio displays four Soft Keys near the middle of the screen.

Figure 35: The four upper Soft Keys

> **ATT**

The ATT Soft Key enables the 6 dB and/or 12 dB front-end attenuators. You can use each attenuator individually or combine them. Select, OFF, 6 dB, 12 dB, or 18 dB.

Tip: hardly anyone uses the ATT control on their HF radio, but they can really improve the signal to noise ratio, especially on the noisy 40m, 80m, and 160m bands. If signals are strong, adding attenuation will make the signal better by cutting out a lot of the background noise. Try it. You will like it!

You can turn the attenuators on and still have AMP1 or AMP2 enabled, but it is preferable to select IPO before inserting an attenuator. Otherwise, the attenuators are simply counteracting the gain of the preamplifiers.

TIP: the 6 dB and 12 dB attenuators on my radio were accurate when tested on the 20m band at an S9 reference with the IPO setting.

The S meter decreases by two S points per 6 dB step. As you would expect, activating both AMP2 and the 18 dB attenuator results in very close to the same S meter reading as selecting IPO and no attenuators.

➢ **IPO**

The IPO Soft Key cycles through three settings. AMP1 (a 10 dB preamplifier which is always in the circuit), AMP2 (which adds 8-10 dB of gain), or IPO (which inserts a 10 dB attenuator to cancel the gain of AMP1).

The 'normal' setting is to leave AMP1 turned on. AMP2 is often used on the quieter 10m and 6m bands. You might consider using the IPO option for the noisier 40m, 60m, 80m, and 160m bands. The IPO setting is remembered when you change bands or switch the radio off.

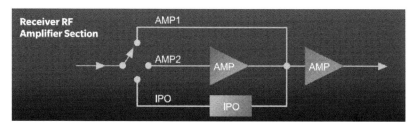

Figure 36: IPO, AMP1 and AMP2 switching, (image Yaesu.com)

According to the Yaesu manual, *"The IPO (Intercept Point Optimization) function maximizes the dynamic range and enhances the close multi-signal and intermodulation characteristics of the receiver."* That sounds mighty impressive, and I expect, like me, you are expecting some "high-tech wizardry." Sadly, all the IPO Soft Key does is remove the AMP2 preamplifier and add around 10 dB of attenuation to the signal. The claim about improving the dynamic range and intermodulation performance is undoubtedly true, but not as exciting as I had expected.

Reducing the input level with an attenuator typically increases the dynamic range and signal to noise ratio and turning off the preamplifiers improves the IMD (intermodulation distortion) performance.

If the received noise level is high, Yaesu recommends selecting the IPO mode, and if the problem persists, adding in some attenuation with the ATT Soft Key. You can also try the noise blanker (NB) and digital noise reduction (DNR).

TIP: On my radio, AMP1 has 10 dB of gain above the IPO setting with an S9 input signal on SSB or CW. Adding AMP2 adds another 8 dB relative to the IPO setting. The S meter increases by 10 dB and 18 dB respectively. Activating both AMP2 and the 18 dB attenuator results in close to the same S meter reading as selecting IPO with no attenuators.

➤ **R.FIL**

The receiver's selectivity and 3rd order IMD dynamic range is greatly improved by the sharp narrowband crystal roofing filters working at the 9 MHz intermediate frequency. They can be selected with the R.FIL Soft Key. Generally, you will use the 500 Hz filter for CW and the 3 kHz filter for SSB. The setting will change to the last used bandwidth if you change between CW and SSB.

The FTDX10 has 500 Hz, 3 kHz, and 12 kHz roofing filters fitted as standard. A 300 Hz filter (XF-130CN) can be added as a user-installed option.

➤ **AGC**

The AGC can be set to OFF (not recommended, even for digital modes), AUTO (default), FAST, MID, or SLOW.

This radio is a hybrid SDR so there should be no need to turn the AGC off for digital modes. In fact, doing so can cause receiver overloading resulting in distortion and poor decodes.

The AGC delay can be set for each of the manual AGC settings, for each mode. You use the FUNC knob to adjust the 'Fast', 'Mid' or 'Slow' delay individually for SSB, AM, FM, PSK & DATA, RTTY, and CW modes.

<FUNC> <RADIO SETTING> <select the mode> <AGC FAST DELAY>

<FUNC> <RADIO SETTING> <select the mode> <AGC MID DELAY>

<FUNC> <RADIO SETTING> <select the mode> <AGC SLOW DELAY>

The CW settings are in the <FUNC> <CW SETTING> <MODE CW> menu.

Upper Soft Keys				
Setting	Options	FTDX10 default settings	ZL3DW settings	My Settings
ATT	OFF, 6 dB, 12 dB, 18 dB	OFF	OFF	
IPO	IPO, AMP1, AMP2	AMP1	AMP1	
R.FIL	500 Hz, 3 kHz, 12 kHz, 300 Hz (option)	3 kHz for SSB 500 Hz for CW	3 kHz for SSB 500 Hz for CW	
AGC	OFF, AUTO, FAST, MID, SLOW	AUTO	AUTO	

Function menu

The menu options and radio settings are accessed by pressing the FUNC knob. In many cases selecting a menu item simply associates the control to the FUNC knob. On the following tables, my settings are in brackets.

LINE 1: SPECTRUM AND WATERFALL

LEVEL -5.0dB	PEAK LVL2	MARKER ON	COLOR 3	CONTRAST 10	DIMMER 10	M-GROUP
MIC GAIN 30	MIC EQ OFF	PROC LEVEL OFF	AMC LEVEL 60	VOX GAIN 10	VOX DELAY 500ms	ANTI VOX 0
RF POWER 100W	MONI LEVEL 35	KEYER ON	BK-IN ON	CW SPEED 22wpm	CW PITCH 750Hz	BK-IN DELAY 200ms
DNF OFF	MESAGE	RECORD	PLAY	DECODE		

Figure 37: The FUNC menu – line 1

Item	Function	Behavior
LEVEL	Sets the spectrum scope level. The spectrum level changes if you change bands, depending on how noisy or active each band is.	The screen disappears but leaves LEVEL associated with the FUNC control. If after adjusting the spectrum level you are not happy with the waterfall brightness, adjust PEAK.
PEAK	Sets the sensitivity of the waterfall display. There are five levels. (LVL 2).	The screen disappears but leaves PEAK associated with the FUNC control. See LEVEL (above).
MARKER	Turns on or off the marker that indicates the receive (green) and transmit (red) VFO frequencies on the spectrum and waterfall display. (ON).	The screen disappears but toggles the marker on or off. I recommend leaving the markers turned on.
COLOR	Allows you to choose from eleven spectrum and waterfall colors.	Brings up the color selecting dialogue box. You have to act fast! It only stays visible for four seconds.

CONTRAST	Sets the display contrast.	The screen disappears but leaves CONTRAST associated with the FUNC control.
DIMMER	Sets the display brightness. Note there is a separate Display Setting for LED brightness.	The screen disappears but leaves DIMMER associated with the FUNC control.
M-GROUP	If 'Groups' are turned on, this item lets you choose a memory channel group.	The screen disappears but leaves M-GROUP associated with the FUNC control.

TIP: there is no dedicated control for setting the spectrum level, which is a shame because you adjust it a lot. If you select LEVEL via the FUNC menu, the LEVEL function will stay allocated to the FUNC knob until you make another selection. This makes it handy for adjusting the spectrum level. Or you can allocate the level adjustment to the VFO tuning ring by press and holding the CS button and selecting LEVEL. I find this the best CS setting.

NOTES:

LEVEL, PEAK, CONTRAST, DIMMER, and M-GROUP can be allocated to the VFO tuning ring by press and holding the CS button. The LEVEL and PEAK functions work on the 3DSS display and the 2D display.

Tip: the LEVEL control also works when the radio is transmitting. It can be used to calibrate the spectrum display to the PEP power level. The spectrum scope is meaningless without a reference. See page 180 for information on how to calibrate the spectrum scope for transmitting. Sadly, there is no separate LEVEL setting for transmitting.

The 3DSS display shares the same color selection as the waterfall and has the same sensitivity. The 2D and 3DSS spectrum displays both have a scale of 5 dB per division. Providing 50 dB of dynamic range.

If you use the AMP1 setting and crank the spectrum level up to +30, the spectrum display can show very weak -130 dBm signals. This is close to the MDS sensitivity. You should be able to see any signal that you can hear over the average received noise level. Normally the received noise is far higher than the MDS sensitivity. It is typically around -103 dBm in an urban environment and -109 dBm in a rural environment. Depending on your location and antenna(s), it may be higher or lower at your place.

On the AMP1 setting with the spectrum level set to zero, the spectrum display will show received signals between about -102 dBm and -52 dBm.

The standard spectrum display has no averaging function. It is always filled and cannot be set to a line or peak-hold display.

LINE 2: TRANSMITTER AUDIO LEVELS

This group contains the microphone adjustments and VOX controls. All of the adjustable controls. MIC GAIN, PROC GAIN, AMC LEVEL, VOX GAIN, DELAY, or ANTI-VOX can be allocated to the VFO tuning ring via the CS button.

Figure 38: The FUNC menu – line 2

Item	Function	Behavior
MIC GAIN	Sets the microphone gain. Measured with the ALC meter. (30).	The screen disappears but leaves MIC GAIN associated with the FUNC control.
MIC EQ[1]	Toggles the parametric microphone equalizer on or off. See note 1 below.	The screen disappears but toggles the parametric microphone equalizer on or off. The control is only active for voice modes.
PROC LEVEL	Turns on the speech processor and sets the compression level. (OFF or 12).	The screen disappears but leaves PROC LEVEL associated with the FUNC control.
AMC LEVEL	Sets the automatic microphone gain control, which is an audio limiter designed to prevent overmodulation of the radio. You can turn it off, but I recommend using this control. (60).	The screen disappears but leaves AMC LEVEL associated with the FUNC control. The control is adjusted for a reading on the COMP meter measured with the speech processor turned off.
VOX GAIN	Allows you to set the VOX gain using the FUNC knob. See more about VOX on page 76. (10)	The screen disappears but leaves VOX GAIN associated with the FUNC control.

VOX DELAY	Allows you to set the VOX delay using the FUNC knob. (500ms)	The screen disappears but leaves VOX DELAY associated with the FUNC control.
ANTI VOX	Allows you to set the anti-VOX protection using the FUNC knob. (0)	The screen disappears but leaves ANTI-VOX associated with the FUNC control.

Note 1: The parametric microphone equalizer (EQ) is a three-stage equalizer that can boost audio frequencies by up to 10 dB or cut audio frequencies by up to 20 dB. It is highly configurable. You can adjust the center frequency of each EQ stage, the boost or cut in dB, and the Q or bandwidth of the filter at each stage.

There are two completely separate equalizers. One for when the speech processor is on and one for when the speech processor is off. See setting up instructions for the parametric equalizer on page 14.

LINE 3: POWER AND CW

This group contains Soft Keys to set the RF Power and Transmit Monitor. They also turn on the CW keyer and Break-in and set the CW speed and pitch.

Figure 39: The FUNC menu – line 3

The RF POWER, MONI LEVEL, CW SPEED, CW PITCH, or the Semi BK-IN DELAY, can be allocated to the VFO tuning ring by press and holding the CS button.

Item	Function	Behavior
RF POWER	Sets the transmitter power between 5W and 100W for the HF bands and the 6m band. (25W for AM). Or 5-50W for the 70 MHz band. (100W).	The screen disappears but leaves RF POWER associated with the FUNC control.

MONI LEVEL	Sets the monitor level that you hear in the speaker or headphones while transmitting. (35). It also independently sets the CW sidetone level.	The screen disappears but leaves MONI LEVEL associated with the FUNC control.
KEYER	Toggles the CW KEYER on or off.	Only active in the CW mode. Does not cause the menu screen to disappear.
BK-IN	Turns Break-in on or off. BK-IN must be on to send voice messages or CW.	Only active in the CW mode. Does not cause the menu screen to disappear.
CW SPEED	Sets the CW keyer speed from 4 to 60 wpm.	The screen disappears but leaves CW SPEED associated with the FUNC control.
CW PITCH	Adjusts the pitch offset for receiving and sending CW. (750 Hz)	The screen disappears but leaves CW PITCH associated with the FUNC control.
BK-IN DELAY[1]	Sets the delay after sending, before the radio switches back to receiving in the Semi Break-in mode. Adjust from 3 ms – 3 secs.	The screen disappears but leaves MONI LEVEL associated with the FUNC control.

Note 1: <FUNC> <BK-IN DELAY> sets the semi break-in delay. That is the time after sending a character before the radio switches back to receiving. The QSK DELAY TIME sets the delay before the transmitter starts to send the CW signal.

LINE 4: DNF, RECORDING, AND DECODE

This group contains Soft Keys to start the CW/PSK/RTTY decoder screen, turn on the digital noise filter (DNR), or carry out an off-air recording.

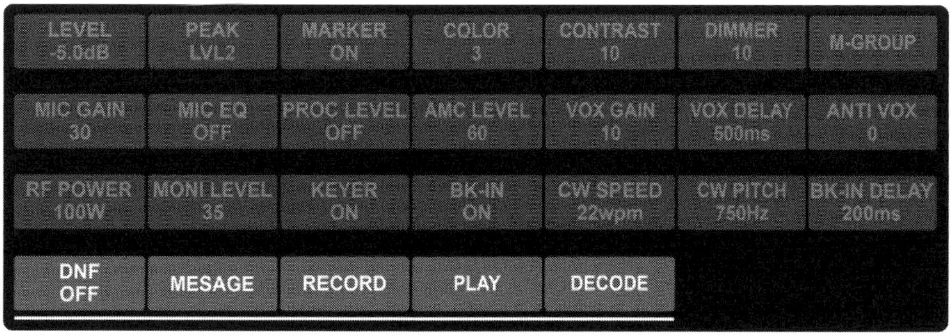

Figure 40: The FUNC menu – line 4

Item	Function	Behavior
DNF[1]	The Digital Notch Filter can eliminate multiple interfering carrier signals. It can even track an interfering signal if it is moving slowly.	Toggles the DNF on or off. Does not cause the menu screen to disappear.
MESSAGE	Opens the message keyer. Voice, RTTY, or PSK depending on the selected mode. Voice on page 16, CW on page 17, PSK/RTTY on page 22	The menu screen disappears, and the MESSAGE MEMORY screen opens.
RECORD	Starts a recording of the receiver audio. An S.REC icon flashes below the MODE indicator. The RECORD Soft Key changes to STOP and recording continues until you touch STOP.	Does not cause the menu screen to disappear. But you can close the menu and the recording will continue. During recording the RECORD Soft Key changes to STOP. You must have an SD card in the radio, to store the recorded file(s).
PLAY	Opens a PLAY LIST with standard 'tape recorder' controls. Select a file and touch the play icon. RX LEVEL changes the playback volume. 20% seems good.	Closes the menu screen and opens a PLAY LIST with standard 'tape recorder' controls. Select a file and play or delete the file. The files are stored as .wav files in the PlayList subdirectory on the SD card.
DECODE	Opens the Decoder window for the internal CW, PSK, or RTTY modes.	The screen disappears and the 'Decode' screen opens. It overlays the spectrum and waterfall display.

Note 1: When you activate the DNF you will be able to see the offending signal disappear from the filter function display, but you might still see it on the spectrum and waterfall display because that shows the incoming RF signal, well before the digital notch filter in the 24 kHz DSP stage.

Radio settings

The Radio Setting menu, <FUNC> <RADIO SETTING>, contains configuration settings for the SSB, AM, FM, PSK, RTTY, and DATA modes.

RADIO SETTING	CW SETTING	OPERATION SETTING	DISPLAY SETTING	EXTENSION SETTING	BACK

TIP: If you make changes to any of the menu settings, it is a good idea to head to <FUNC> <EXTENSION SETTING> <SD CARD> <Menu Save DONE> <NEW> <ENT> and save your changes to the SD card.

SSB MODE

The first three settings are the AF (audio frequency) **Treble, Middle, and Bass** tone controls. The default is for all three to be set to zero creating a flat audio response, but you can boost any of them by as much as 10 dB or cut them by up to 20 dB. This adjustment is great if you have a hearing impairment, or just like the radio to sound different.

Note: this is a plus for the FTDX10. The FTDX101 lacks any receiver tone controls.

The **AGC Fast, Mid, or Slow delay** settings control how fast the AGC action will decay after the received signal falls below the AGC threshold if manual AGC has been selected. They have no effect if the AGC is set to AUTO.

AGC FAST DELAY can be adjusted from 20 – 4000 ms, (default = 300 ms).

AGC MID DELAY can be adjusted from 20 – 4000 ms, (default = 1000 ms).

AGC SLOW DELAY can be adjusted from 20 – 4000 ms, (default = 3000 ms).

TIP: AUTO, FAST, MID, or SLOW AGC can be selected for each band. Touch the AGC Soft Key on the bar that separates the VFO numbers from the spectrum scope or 3DSS display. Yaesu recommends that you leave it set to AUTO which selects the optimum AGC setting for the selected mode.

Receiver audio filter: The next four settings control the shape of the receiver audio passband. I left all these settings at the default settings. LCUT FREQ sets the low-frequency end of the passband and LCUT SLOPE adjusts how sharp the filter is. 6 dB per octave is a soft filter and 18 dB per octave is a sharp filter. You would normally make both ends sharp or soft, not mix them. Naturally, HCUT FREQ sets the high-frequency end of the passband and HCUT SLOPE adjusts how sharp the filter roll-off is at the high end. I recommend the sharper 18 dB per octave setting for contesting because the greater selectivity will help you to reject adjacent signals.

SSB receiver output: SSB OUT LEVEL sets the level of the audio signal to the RTTY/DATA jack. 0 – 50 -100. This setting is only useful if you are connecting an external device "the old-fashioned way." My connections to the PC are over the single USB cable, so the setting is unimportant.

Transmitter bandwidth: TX BPF SEL selects the bandwidth of the transmitted signal. The default for SSB is 100-2900 Hz (2.8 kHz BW). If you are contesting you may prefer a narrow option such as 200-2800 Hz (2.6 kHz), or 300 to 2700 Hz (2.4 kHz).

SSB mod source is a bit "oddball." It can be set to MIC in which case the SSB modulation comes from the microphone and not the RTTY/Data jack on the rear panel or the USB cable. "So far so good." If you set it to REAR, the modulating audio comes from the rear panel if you key the transmitter with the MOX switch. However, the microphone behaves normally if you use the PTT switch on the hand microphone or key the radio using the PTT jack on the rear panel. I guess that makes sense if you are using the rear panel PTT for a foot switch.

TIP: You would only select REAR if you were planning on sending speech audio to the transceiver from another device. So that the equalizer and compressor settings would be active. You should use the DATA-U mode for connecting external digital mode software. 'That's what it's for!' My advice is to set SSB MOD SOURCE to MIC unless you have a good reason not to.

Rear select: If you selected REAR as the SSB MOD SOURCE, then REAR SELECT sets whether the SSB modulation audio will come from the RTTY/Data jack or the USB cable.

RPORT gain sets the audio level into the transmitter on SSB mode. Assuming you took my advice and set SSB MOD SOURCE to MIC, then this control is irrelevant for the SSB mode. In any case, I would leave it at 50 initially and adjust the modulation level using the PC (Windows) sound card levels and/or the level control on the external software or connected device.

RPTT select sets the control line for PTT, ('keying' the transmitter). Set it to RTS (ready to send) for the USB port, or DAKY if you are using pin 3 of the RTTY/Data jack for PTT.

TIP: The RTS and DTR labels don't matter. You can use either line for the transmit PTT as long as you use the other line for CW or RTTY. The names relate to old-fashioned RS-232 communications between 'old school' computers. RTS stands for 'ready to send' and DTR stands for 'data terminal ready.' But the lines have not been used for that sort of signaling since the 1970s.

RADIO SETTINGS - SSB			
Function	Range	ZL3DW Setting	My Setting
AF TREBLE GAIN	-20 dB to +10 dB	0 (default)	
AF MIDDLE GAIN	-20 dB to +10 dB	0 (default)	
AF BASS GAIN	-20 dB to +10 dB	0 (default)	
AGC FAST DELAY	20-4000 ms	300 ms (default)	
AGC MID DELAY	20-4000 ms	1000 ms (default)	
AGC SLOW DELAY	20-4000 ms	3000 ms (default)	
LCUT FREQ	Off / 100 – 1000 Hz	100 Hz (default)	
LCUT SLOPE	6 or 18 dB/octave	18 dB/oct	
HCUT FREQ	Off / 700-4000 Hz	3000 Hz (default)	
HCUT SLOPE	6 or 18 dB/octave	18 dB/oct	
SSB OUT LEVEL	0-100	50 (default)	
TX BPF SEL	50-3050, 100-2900, 200-2800, 300-2700, 400-2600 Hz	100-2900 Hz (default)	
SSB MOD SOURCE	MIC/REAR	MIC (default)	
REAR SELECT	DATA / USB	DATA (default)	
RPORT GAIN	0-100	50 (default)	
RPTT SELECT	DAKY/RTS/DTR	DAKY (default)	

AM MODE

The first three settings are the AF (audio frequency) **Treble, Middle, and Bass** tone controls. The default is for all three to be set to zero creating a flat audio response, but you can boost any of them by as much as 10 dB or cut them by up to 20 dB. This adjustment is great if you have a hearing impairment, or just like the radio to sound different. Some Bass boost may be nice if you are listening to shortwave AM.

The **AGC Fast, Mid, or Slow delay** settings control how fast the AGC action will decay after the received signal falls below the AGC threshold if manual AGC has been selected. They have no effect if the AGC is set to AUTO.

AGC FAST DELAY can be adjusted from 20 – 4000 ms, (default = 1000 ms).

AGC MID DELAY can be adjusted from 20 – 4000 ms, (default = 2000 ms).

AGC SLOW DELAY can be adjusted from 20 – 4000 ms, (default = 4000 ms).

TIP: AUTO, FAST, MID, or SLOW AGC can be selected for each band. Touch the AGC Soft Key on the bar that separates the VFO numbers from the spectrum scope or 3DSS display. Yaesu recommends that you leave it set to AUTO which selects the optimum AGC setting for the selected mode and makes these delay settings irrelevant.

Receiver audio filter: The next four settings control the shape of the receiver audio passband. The default is to have the filters turned OFF so that you hear the full bandwidth on shortwave AM. You should set the roofing filter R.FIL to 12 kHz as well.

If you are listening to amateur radio signals on AM, set the LCUT FREQ to 100 Hz, and the HCUT FREQ somewhere between 3000 and 4000 Hz. I recommend using the default 6 dB per octave setting for the two SLOPE settings.

LCUT FREQ sets the low-frequency end of the passband and LCUT SLOPE adjusts how sharp the filter is. 6 dB per octave is a soft filter and 18 dB per octave is a sharp filter. You would normally make both ends sharp or soft, not mix them. Naturally, HCUT FREQ sets the high-frequency end of the passband and HCUT SLOPE adjusts how sharp the filter roll-off is at the high end.

AM receiver output: AM OUT LEVEL sets the level of the audio signal to the RTTY/DATA jack. 0 – 50 -100. This setting is only useful if you are connecting an external device "the old-fashioned way." My connections to the PC are over the single USB cable, so the setting is unimportant.

Transmitter bandwidth: TX BPF SEL selects the bandwidth of the transmitted AM signal. The default is 50-3050 Hz (3 kHz BW). I can't see any reason to change it.

Audio inputs: AM MOD SOURCE. This one is a bit "oddball." It can be set to MIC in which case the AM modulation comes from the microphone and not the RTTY/Data jack on the rear panel or the USB cable. "So far so good." If you set it to REAR, the modulating audio comes from the rear panel if you key the transmitter with the MOX switch. However, the microphone behaves normally if you use the PTT switch on the hand microphone or key the radio using the PTT jack on the rear panel. If you selected REAR as the AM MOD SOURCE, then **REAR SELECT** sets whether the AM modulation audio will come from the RTTY/Data jack or the USB cable.

TIP: You would only select REAR if you were planning on sending speech audio to the transceiver from another device. So that the equalizer and compressor settings would be active. My advice is to set AM MOD SOURCE to MIC.

Mic gain sets the microphone gain for AM. You can set it anywhere from 1 to 100 using the FUNC knob. After that, you should not have to adjust it again unless you change microphones. Or you can leave it on the default MCVR setting, which uses the same <FUNC> <MIC GAIN> as you set for SSB.

TIP: changing the <FUNC> <MIC GAIN> affects the microphone level for SSB, AM, and FM. I recommend setting it for SSB and then leaving it alone. If you want a different microphone sensitivity for AM, use one of the 100 pre-set levels.

NOTE: The manual refers to using the MIC/SPEED knob on the front panel. This is a cut-n-paste error from the FTDX101 manual. The FTDX10 does not have a MIC/SPEED knob.

Rear select: If you selected REAR as the SSB MOD SOURCE, then REAR SELECT sets whether the AM modulation audio will come from the RTTY/Data jack or the USB cable.

RPORT gain sets the audio level into the transmitter from the USB port or RTTY/Data jack on AM mode. I would leave it at 50 and adjust the modulation level using the PC (Windows) sound card levels and/or the level control on the external software or connected device.

RPTT select sets the control line for PTT, ('keying' the transmitter). Set it to RTS (ready to send) for the USB port, or DAKY if you are using pin 3 of the RTTY/Data jack for PTT.

TIP: The RTS and DTR labels don't matter. You can use either line for the transmit PTT as long as you use the other line for CW or RTTY. The names relate to old-fashioned RS-232 communications between 'old school' computers. RTS stands for 'ready to send' and DTR stands for 'data terminal ready.' But the lines have not been used for that sort of signaling since the 1970s. Anyway, I always use 'ready to send' for the 'Send' or 'PTT' command and 'data terminal ready' for the CW or RTTY data signal.

RADIO SETTINGS - AM			
Function	Range	ZL3DW Setting	My Setting
AF TREBLE GAIN	-20 dB to +10 dB	0 (default)	
AF MIDDLE GAIN	-20 dB to +10 dB	0 (default)	
AF BASS GAIN	-20 dB to +10 dB	0 (default)	
AGC FAST DELAY	20-4000 ms	1000 ms (default)	
AGC MID DELAY	20-4000 ms	2000 ms (default)	
AGC SLOW DELAY	20-4000 ms	4000 ms (default)	
LCUT FREQ	Off/100 – 1000 Hz	OFF (default)	
LCUT SLOPE	6 or 18 dB/octave	6 dB/oct (default)	
HCUT FREQ	700-4000 Hz	OFF (default)	
HCUT SLOPE	6 or 18 dB/octave	6 dB/oct (default)	
AM OUT LVL	0-100	50 (default)	
TX BPF SEL	50-3050, 100-2900, 200-2800, 300-2700, 400-2600 Hz	50-3050 Hz (default)	
AM MOD SOURCE	MIC/REAR	MIC (default)	
MIC GAIN	MCVR / 0-100	MCVR (default)	
REAR SELECT	DATA/USB	DATA (default)	
RPORT GAIN	0-100	50 (default)	
RPTT SEL	DAKY/RTS/DTR	DAKY (default)	

FM MODE

The first three settings are the AF (audio frequency) **Treble, Middle, and Bass** tone controls. The default is for all three to be set to zero creating a flat audio response, but you can boost any of them by as much as 10 dB or cut them by up to 20 dB. This adjustment is great if you have a hearing impairment, or just like the radio to sound different.

The **AGC Fast, Mid, or Slow delay** settings control how fast the AGC action will decay after the received signal falls below the AGC threshold if manual AGC has been selected. They have no effect if the AGC is set to AUTO.

AGC FAST DELAY can be adjusted from 20 – 4000 ms, (default = 160 ms).

AGC MID DELAY can be adjusted from 20 – 4000 ms, (default = 500 ms).

AGC SLOW DELAY can be adjusted from 20 – 4000 ms, (default = 1500 ms).

TIP: The AGC can be selected for each band. It is a Soft Key on the bar that separates the VFO from the spectrum scope or 3DSS display. Yaesu recommends that you leave it set to Auto which makes the AGC decay settings irrelevant.

Receiver audio filter: The next four settings control the shape of the receiver audio passband. I left all these settings at the default settings. LCUT FREQ sets the low-frequency end of the passband and LCUT SLOPE adjusts how sharp the filter is. 6 dB per octave is a soft filter and 18 dB per octave is a sharp filter. The default for FM is to use the sharp filter. You would normally make both ends sharp or soft, not mix them. HCUT FREQ sets the high-frequency end of the passband and HCUT SLOPE adjusts how sharp the filter roll-off is at the high end.

FM receiver output: FM OUT LEVEL sets the level of the audio signal to the RTTY/DATA jack. These settings are only useful if you are connecting up an external device "the old-fashioned way." All My connections to the PC are over the USB cable, so this setting is unimportant.

Audio inputs: FM MOD SOURCE. This one is a bit "oddball." It can be set to MIC in which case the FM modulation comes from the microphone and not the RTTY/Data jack on the rear panel or the USB cable. "So far so good." If you set it to REAR, the modulating audio comes from the rear panel if you key the transmitter with the MOX switch. However, the microphone behaves normally if you use the PTT switch on the hand microphone or key the radio using the PTT jack on the rear panel. If you did select REAR as the FM MOD SOURCE, then REAR SELECT sets whether the AM modulation audio will come from the RTTY/Data jack or the USB cable. *TIP: You would only select REAR if you were planning on sending speech audio to the transceiver from another device. So that the equalizer and compressor settings would be active. My advice is to set FM MOD SOURCE to MIC.*

Mic gain sets the microphone gain for FM. You can set it anywhere from 1 to 100 using the FUNC knob. After that, you should not have to adjust it again unless you change microphones. Or you can leave it on the default MCVR setting, which uses the same <FUNC> <MIC GAIN> as you set for SSB.

TIP: Changing the <FUNC> <MIC GAIN> affects the microphone level for SSB, AM, and FM. I recommend setting it for SSB and then leaving it alone. If you want a different microphone sensitivity for AM, use one of the 100 pre-set levels.

NOTE: The manual refers to using the MIC/SPEED knob on the front panel. This is a cut-n-paste error from the FTDX101 manual. The FTDX10 does not have a MIC/SPEED knob.

Rear select: If you selected REAR as the SSB MOD SOURCE, then REAR SELECT sets whether the FM modulation audio will come from the RTTY/Data jack or the USB cable.

RPORT gain sets the audio level into the transmitter from the USB port or RTTY/Data jack on FM mode. I would leave it at 50 and adjust the modulation level using the PC (Windows) sound card levels and/or the level control on the external software or connected device.

RPTT select sets the control line for PTT, ('keying' the transmitter). Set it to RTS (ready to send) for the USB port, or DAKY if you are using pin 3 of the RTTY/Data jack for PTT.

➢ **FM Repeaters:**

You can use **RPT SHIFT (28 MHz)** to change the offset between your receiver and transmitter frequencies for operating through an FM repeater on the 10m band. The default 100 kHz offset suits most repeaters in most countries.

You can use **RPT SHIFT (50 MHz)** to change the offset between your receiver and transmitter frequencies for operating through an FM repeater on the 6m band. The default 1 MHz offset suits the repeaters in most countries. But some 6m repeaters use a 500 kHz offset.

The **RPT** setting applies the specified repeater shift. SIMP (default) means simplex. The normal mode for the transceiver.

NOTE: The repeater settings are saved for FM on the currently selected band.

With a plus (+) offset the transmitter will transmit higher than the receive frequency. With a minus (-) offset the transmitter will transmit lower than the receive frequency.

TONE FREQ sets the CTCSS tone frequency. See 'Setting the Tone' on page 89.

ENC/DEC turns on the tone-coded squelch system.

- OFF, no tone is transmitted.

- ENC, a tone is transmitted to open the repeater squelch.

- TSQ, a tone is transmitted to open the repeater squelch and the same tone must be received to open the squelch on the FTDX10.

RADIO SETTINGS - FM			
Function	Range	ZL3DW Setting	My Setting
AF TREBLE GAIN	-20 dB to +10 dB	0 (default)	
AF MIDDLE GAIN	-20 dB to +10 dB	0 (default)	
AF BASS GAIN	-20 dB to +10 dB	0 (default)	
AGC FAST DELAY	20-4000 ms	160 ms (default)	
AGC MID DELAY	20-4000 ms	500 ms (default)	
AGC SLOW DELAY	20-4000 ms	1500 ms (default)	
LCUT FREQ	Off/100 – 1000 Hz	300 (default)	
LCUT SLOPE	6 or 18 dB/octave	18 dB/oct (default)	
HCUT FREQ	700-4000 Hz	3000 (default)	
HCUT SLOPE	6 or 18 dB/octave	18 dB/oct (default)	
FM OUT LVL	0-100	50 (default)	
FM MOD SOURCE	MIC/REAR	MIC (default)	
MIC GAIN	MCVR / 0-100	MCVR (default)	
REAR SELECT	DATA/USB	DATA (default)	
RPORT GAIN	0-100	50 (default)	
RPTT SEL	DAKY/RTS/DTR	DAKY (default)	
RPT SHIFT (28 MHz)	0-1000 kHz	100 kHz (default)	
RPT SHIFT (50 MHz)	0-4000 kHz	1000 kHz (default)	
RPT (repeater offset)	SIMP + -	SIMP (default)	
TONE FREQ	CTCSS table p89.	67.0 (default)	
ENC/DEC	OFF/ENC/TSQ	OFF	

PSK AND DATA MODES

These settings are for the PSK mode and the DATA modes. Most digital mode software developers expect that you will use the upper sideband DATA-U mode on all bands.

The first three settings are the AF (audio frequency) **Treble, Middle, and Bass** tone controls. The default is for all three to be set to zero creating a flat audio response. You would not normally want to introduce audio slope for digital modes.

The **AGC Fast, Mid, or Slow delay** settings control how fast the AGC action will decay after the received signal falls below the AGC threshold if manual AGC has been selected. They have no effect if the AGC is set to AUTO.

AGC FAST DELAY can be adjusted from 20 – 160 ms, (default = 300 ms).

AGC MID DELAY can be adjusted from 20 – 500 ms, (default = 1000 ms).

AGC SLOW DELAY can be adjusted from 20 – 1500 ms, (default = 3000 ms).

TIP: AUTO, FAST, MID, or SLOW AGC can be selected for each band. Touch the AGC Soft Key on the bar that separates the VFO numbers from the spectrum scope or 3DSS display. Yaesu recommends that you leave it set to AUTO which selects the optimum AGC setting for the selected mode.

This radio is a hybrid SDR so there should be no need to turn the AGC off for digital modes. Doing so could cause receiver overloading resulting in distortion and poor decodes.

PSK and Data mode offsets: The manual does not really explain what PSK TONE does. It sets the carrier offset for a PSK signal when the radio is in the PSK mode. It is rather like the CW pitch setting. At the displayed VFO frequency a received PSK signal will sound like a 1 kHz (default), 1.5 kHz, or 2 kHz signal.

The **data shift (SSB)** setting does the same thing for the DATA_U (or DATA-L) modes. It sets the frequency offset above (or below) the carrier frequency that a data signal will be placed. It is the offset that will apply for FSK RTTY or 300 baud HF packet radio. You can adjust the offset from 0 to 3 kHz in 10 Hz steps. The default is 1500 Hz. I can't see any reason to change it.

Receiver audio filter: The next four settings control the shape of the receiver audio passband. I left all these settings at the default settings. LCUT FREQ sets the low-frequency end of the passband and LCUT SLOPE adjusts how sharp the filter is. 6 dB per octave is a soft filter and 18 dB per octave is a sharp filter. The default for PSK/DATA is to use the sharp filter. You would normally make both ends sharp or soft, not mix them. Naturally, HCUT FREQ sets the high-frequency end of the passband and HCUT SLOPE adjusts how sharp the filter roll-off is at the high end.

PSK or DATA receiver output: DATA OUT LEVEL sets the level of the audio signal to the RTTY/DATA jack. 0 – 50 -100. These settings are only useful if you are connecting an external device "the old-fashioned way." My connections to the PC are over the USB cable, so this setting is unimportant.

Transmitter bandwidth: TX BPF SEL selects the bandwidth of the transmitted signal. I changed the DATA mode bandwidth to 100-2900 Hz (2.8 kHz BW) because it almost matches the bandwidth of the 200-3000 Hz receiver passband displayed on my digital mode software. The 50-3050 Hz option may fit better, but I would not be comfortable transmitting a digital mode signal below 100 Hz. The default is 300-2700 Hz (2.4 kHz). Fine for FT8, but a little narrow for external PSK or RTTY.

Audio inputs: DATA MOD SOURCE. Needs to be set to REAR so that you can send data from an external digital modes program over the USB cable. REAR SELECT should be set to USB unless you are using a Rig Runner or old-fashioned audio cables.

RPORT gain sets the audio level into the transmitter from the USB port or RTTY/Data jack on DATA mode. You can leave it at 12-15 if you like and adjust the modulation level using the PC (Windows) sound card levels and/or the level control on the external software or connected device. The applied audio level should be adjusted so that the transmitter just reaches full power or a little less. The ALC meter should be low or zero for digital modes like PSK, RTTY, or FT8. It should remain within the white area for speech.

TIP: it is better to have the transceiver setting high and the Windows setting low. Rather than the other way around. This reduces the risk of audio harmonics causing a second instance of your digital mode being transmitted along with the wanted signal. Transmitting on two frequencies at the same time causes all sorts of confusion and can make you very unpopular. I set my PSK/DATA RPORT GAIN to 12.

RPTT select sets the control line for activating 'keying' the transmitter from the USB cable. Set it to RTS (ready to send), or DAKY if you are using pin 3 of the RTTY/Data jack for PTT.

RADIO SETTINGS – PSK and DATA			
Function	Range	ZL3DW Setting	My Setting
AF TREBLE GAIN	-20 dB to +10 dB	0 (default)	
AF MIDDLE GAIN	-20 dB to +10 dB	0 (default)	
AF BASS GAIN	-20 dB to +10 dB	0 (default)	
AGC FAST DELAY	20-4000 ms	160 ms (default)	
AGC MID DELAY	20-4000 ms	500 ms (default)	
AGC SLOW DELAY	20-4000 ms	1500 ms (default)	
PSK TONE	1000/1500/2000	1000 Hz (default)	
DATA SHIFT (SSB)	0-3000 Hz	1500 Hz (default)	
LCUT FREQ	Off / 100 – 1000 Hz	300 Hz (default)	

LCUT SLOPE	6 or 18 dB/octave	18 dB/oct (default)	
HCUT FREQ	Off / 700-4000 Hz	3000 Hz (default)	
HCUT SLOPE	6 or 18 dB/octave	18 dB/oct (default)	
DATA OUT LEVEL	0-100	50 (default)	
TX BPF SEL	50-3050, 100-2900, 200-2800, 300-2700, 400-2600 Hz	100-2900 Hz	
DATA MOD SOURCE	MIC/REAR	REAR	
REAR SEL	DATA/USB	USB	
RPORT GAIN	0-100	12	
RPTT SEL	DAKY/RTS/DTR	RTS	

RTTY MODE

These settings are for the internal RTTY mode. The normal mode for amateur radio RTTY is lower sideband RTTY-L for all bands.

RTTY is a frequency shift keying (FSK) mode. It works by sending a tone on a 'mark' frequency and then applying frequency shifts to a 'space' frequency to send the Baudot code. For amateur radio RTTY, the 'space' frequency is usually 170 Hz lower than the 'mark' frequency. The mark frequency is the frequency indicated by the VFO. It has been offset from the carrier point by the **Mark Frequency** usually 2125 Hz.

TIP: in the built-in RTTY mode, the Mark and Space frequencies are shown as markers on the filter function display.

The first three settings are the AF (audio frequency) **Treble, Middle, and Bass** tone controls. The default is for all three to be set to zero creating a flat audio response. I don't know of any reason why you would want to introduce audio slope for RTTY. But I suppose you could experiment with cutting the treble and bass and/or boosting the mid-range where the RTTY tones lie.

The **AGC Fast, Mid, or Slow delay** settings control how fast the AGC action will decay after the received signal falls below the AGC threshold. FAST, MID, and SLOW refer to the AGC setting that you have selected on the main display. The default settings are the same as the ones chosen for FM, PSK, and DATA.

RTTY polarity: POLARITY RX and POLARITY TX invert the tone frequencies of the RTTY signal. Amateur radio RTTY uses the normal (NOR) mode where the 'space' frequency is lower than the 'mark' frequency. Commercial operators on the shortwave bands often use the reverse (REV) mode where the 'space' frequency is higher than the 'mark' frequency.

Receiver audio filter: The next four settings control the shape of the receiver audio passband. I left all these settings at the default settings.

LCUT FREQ sets the low-frequency end of the passband and LCUT SLOPE adjusts how sharp the filter is. Naturally, HCUT FREQ sets the high-frequency end of the passband and HCUT SLOPE adjusts how sharp the filter roll-off is at the high end. For some reason, Yaesu has used the sharp 18 dB/octave roll off on the low-frequency filter, but the softer 6 dB/octave roll off for the high end of the audio passband. There must be a reason, so I left my radio set to the default settings.

RTTY receiver output: RTTY OUT LEVEL sets the level of the audio signal to the RTTY/DATA jack. 0-50-100. These settings are only useful if you are connecting an external device "the old-fashioned way." My connections to the PC are over the USB cable, so this setting is unimportant.

RPTT select sets the control line for activating 'keying' the transmitter from the USB cable. Set it to RTS (ready to send), or DAKY if you are using pin 3 of the RTTY/Data jack for PTT.

RTTY mark and shift: MARK FREQUENCY sets the audio tone for the mark frequency. The choices are the two internationally recognized 'low tones.' The default 2125 Hz setting is the most used offset, but the alternative setting of 1275 Hz is perfectly acceptable as well. The SHIFT frequency for amateur radio is 170 Hz.

Some people use 200 Hz. The other shift choices are for receiving commercial RTTY transmissions which may use other shift frequencies and baud rates.

RADIO SETTINGS - RTTY			
Function	Range	ZL3DW Setting	My Setting
AF TREBLE GAIN	-20 dB to +10 dB	0 (default)	
AF MIDDLE GAIN	-20 dB to +10 dB	0 (default)	
AF BASS GAIN	-20 dB to +10 dB	0 (default)	
AGC FAST DELAY	20-4000 ms	160 ms (default)	
AGC MID DELAY	20-4000 ms	500 ms (default)	
AGC SLOW DELAY	20-4000 ms	1500 ms (default)	
POLARITY RX	NOR/REV	NOR (default)	
POLARITY TX	NOR/REV	NOR (default)	
LCUT FREQ	1000/1500/2000	300 Hz (default)	
LCUT SLOPE	6 or 18 dB/octave	18 dB/oct (default)	
HCUT FREQ	700-4000 Hz	3000 Hz (default)	
HCUT SLOPE	6 or 18 dB/octave	6 dB/oct (default)	
RTTY OUT LVL	0-100	50 (default)	
RPTT SEL	DAKY/RTS/DTR	RTS	
MARK FREQUENCY	1275 or 2125	2125 Hz (default)	
SHIFT FREQUENCY	170, 200, 425 or 850 Hz	170 Hz (default)	

ENCODING AND DECODING PSK (ENCDEC PSK)

PSK mode: You can choose from BPSK which is standard PSK-31, or QPSK which is four-phase (quadrature) PSK. I assume that it is QPSK-31 since there is no baud rate control. QPSK has the advantage of having built-in error correction, but it is considerably less sensitive than BPSK and much less popular on the bands. The 31 symbols per second baud rate of PSK-31 is chosen to provide a narrow bandwidth transmission at a typical typing speed. These days there is little PSK activity since most of the world has moved to FT8.

The internal PSK decoder has AFC (automatic frequency control) to help you tune the PSK signal accurately. It will pull the receiver slightly to get the signal tuned in. The DECODE AFC RANGE sets the bandwidth for the AFC action. A BPSK-31 signal has a bandwidth of about 31 Hz. So, I would experiment with using the 30 Hz setting or the default 15 Hz setting.

QPSK phase shift: You can change the direction of the QPSK phase shift using the QPSK POLARITY settings. If you can't decode a QPSK signal, it is possible that the other station is using the other mode.

QPSK data streams are not pseudo-random scrambled, but the transceiver introduces a built-in phase rotation to make sure that the overall phase pattern is random. The transmitted phase state will rotate by 90 degrees for each (2-bit) data symbol. The QPSK POLARITY controls set whether the data rotation is 90 degrees clockwise or 90 degrees anti-clockwise. Without this rotation, the transmission would not be random if the transmitted data contained a long string of 1s (or 0s). That would make it difficult (or impossible) to recover the data clock signal from the received transmission. The phase states of the incoming QPSK signals are averaged to lock the decoder clock.

The **PSK TX level** sets the level of PSK that results from a PSK signal applied to the RTTY/DATA jack. It does not affect PSK sent using the MESSAGE keyer, which always transmits at just below full power, peaking to full power at the end of the transmission. The level can be set from 0 to 100. The default setting is 70.

RADIO SETTINGS – ENCODING & DECODING PSK			
Function	Range	ZL3DW Setting	My Setting
PSK MODE	BPSK or QPSK	BPSK (default)	
DECODE AFC RANGE	8, 15, or 30 Hz	15 Hz (default)	
QPSK POLARITY RX	NOR/REV	NOR (default)	
QPSK POLARITY TX	NOR/REV	NOR (default)	
PSK TX LEVEL	0-100	70 (default)	

ENCODING AND DECODING RTTY (ENCDEC RTTY)

These settings relate to FSK RTTY generated by the radio in the RTTY mode.

You can also operate RTTY in the DATA-U mode using AFSK RTTY from an external digital modes program. FSK = frequency shift keying, AFSK = audio frequency shift keying.

USOS: stands for 'unshift on space.' The Baudot code that is used to send RTTY dates back to the old mechanical teletype machine days. It is a 5-bit code. Five bits only allows for 32 different characters, so there are not enough combinations to encode the 26 upper-case letters and numerals 0 to 9. Baudot gets around this by sending a special 'FIGS' code before sending numbers and then sending a 'LTRS' code to switch the decoder back to letters after the numbers have been sent. Problems arise if noise on the signal prevents the FIGS or LTRS code from being received properly. If that happens the decoder outputs a string of garbage characters until the next time a FIGS or LTRS code arrives.

If you have TX USOS turned on, the encoder will send a LTRS code every time a 'space' character is sent. Of course, this happens at the end of every word. These regular reminders that the data is letters, not numbers, greatly improve the data synchronization between the transmitting station and the receiving station, resulting in fewer errors in the text. If you have RX USOS turned on, the decoder in the radio will return to letters mode every time a space character is received regardless of whether a LTRS code has been received from the distant station.

Sending the extra LTRS codes does slow the transmission down a little. The word rate changes from slow to slightly slower, so the benefits outweigh the disadvantages.

The moral of the story is, leave RX USOS and TX USOS turned ON.

RX new line code: The decoded RTTY text is much easier to read if the lines and paragraphs don't run together. This setting starts a new line on the decode screen each time a CR (carriage return), LF (line feed), or combined CR+LF character is received. You can change to only inserting a new line when the combined CR+LF character is received, but I recommend leaving it at the default setting.

TX diddle: When switched to transmit mode, mechanical RTTY machines would send a blank signal at any time that a character was not being sent. It is called a 'diddle' because it sounds like a warble, mostly mark tone with regular transitions to the space frequency. It allowed the receiver at the distant end to stay locked.

If you are a slow typist or you pause between words, the RTTY encoder software will send the diddle signal between words or characters as well. Yaesu has given you the option of sending the standard BLANK signal, or a string of LTRS codes to ensure that the decoder at the distant station stays set for receiving letters, or you can turn the diddle off. In which case, nothing is sent except the text message.

If signals are reasonable, that should be fine. The only reason to send the diddle is to keep the distant station synchronized and tuned to your signal. The default is BLANK.

The final setting is the **Baudot code**. This one is a mystery. I have never seen this choice offered on non-Yaesu transceivers. I believe that the US setting refers to the ITA2 (international telegraph alphabet no 2) alphabet and the CCITT option refers to the 'Murray' code, otherwise known as CCITT2. Apparently, there are very minor differences in the way some of the letters and characters look when printed.

I guess you could try the CCITT version if you are working a European station and see if you can spot any differences.

RADIO SETTINGS – ENCODING & DECODING RTTY			
Function	Range	ZL3DW Setting	My Setting
RX USOS	OFF/ON	ON (default)	
TX USOS	OFF/ON	ON (default)	
RX NEW LINE CODE	CR, LF, CR+LF or just CR+LF	CR, LF, CR+LF (default)	
TX DIDDLE	OFF, BLANK, or LTRS	BLANK (default)	
BAUDOT CODE	CCITT or US	US (default)	

Figure 41: Creed 75 Teleprinter.   Image https://www.worthpoint.com/

Teleprinters like this were used to send messages over telephone wires and radio systems using RTTY. I had one which I used as a computer printer way back in the mid-1970s! Note the LTRS/FIGS toggle key on the right.

CW Settings

The CW Setting menu, <FUNC> <CW SETTING>, includes settings relating to CW operation. Including break-in, QSK, audio frequency response, keyer settings, Morse key type, and the CW decoder bandwidth.

TIP: If you make changes to any of the menu settings, it is a good idea to head to <FUNC> <EXTENSION SETTING> <SD CARD> <Menu Save DONE> <NEW> <ENT> and save your changes to the SD card.

MODE CW

The first three settings are the **AF (audio frequency) Treble, Middle, and Bass** tone controls. The default is for all three to be set to zero creating a flat audio response, but you can boost any of them by as much as 10 dB or cut them by up to 20 dB. This adjustment is great if you have a hearing impairment, or just like the radio to sound different.

The **AGC Fast, Mid, or Slow delay** settings control how fast the AGC action will decay after the received signal falls below the AGC threshold if manual AGC has been selected. They have no effect if the AGC is set to AUTO.

AGC FAST DELAY can be adjusted from 20 – 4000 ms, (default = 160 ms).

AGC MID DELAY can be adjusted from 20 – 4000 ms, (default = 500 ms).

AGC SLOW DELAY can be adjusted from 20 – 4000 ms, (default = 1500 ms).

TIP: The AGC can be selected for each band. It is a Soft Key on the bar that separates the VFO from the spectrum scope or 3DSS display. Yaesu recommends that you leave it set to Auto which makes the AGC decay settings irrelevant.

Receiver audio filter: The next four settings control the shape of the receiver audio passband. I left all four settings at the defaults.

LCUT FREQ sets the low-frequency end of the passband. The default frequency is 250 Hz. The default for CW is to use the sharper 18 dB per octave filter shape.

HCUT FREQ sets the high-frequency end of the passband and HCUT SLOPE adjusts how sharp the filter roll-off is at the high end. The default HCUT frequency is 1200 Hz. It is lower than on the other modes because you normally use a CW pitch of between 700 and 100 Hz.

CW out level sets the level of the audio signal to the RTTY/DATA jack. These settings are only useful if you are connecting up an external device "the old-fashioned way." My connections to the PC are over the USB cable, so this setting is unimportant.

CW auto mode: This lets you enable the Morse key or paddle when you are on SSB. Some people like to use a CW sign-off when operating SSB and there is certainly no prohibition against using CW on the SSB band segment. When set to ON the key is 'live' but as usual CW will only be transmitted if <FUNC> <BK-IN> is enabled. When CW auto mode is set to OFF, the key is disabled on SSB. If you select the 50M option, the key is active on the 6m band but not on the HF bands. I suggest leaving this function turned off unless you really want the ability to send CW in SSB mode.

TIP: You might consider leaving the CW AUTO MODE turned ON and controlling whether the CW will be transmitted by activating BK-IN. This is fine. But note that BK-IN has to be turned on for the voice message keyer to work.

CW break-in type selects SEMI break-in or FULL break-in.

CW wave shape sets the shape of the CW waveform (keying envelope rise and fall times). The default setting is 6 ms. Selecting a slower rise time will make your signal sound softer. Choosing the faster 4 ms rise time will make your signal sound a little harsher. It should only be selected if you are using high-speed CW.

CW frequency display adjusts the offset that will be applied to the VFO when you tune to a CW signal and then change to SSB. The default 'PITCH OFFSET' setting offsets the VFO when you switch from CW to SSB, so you hear the CW signal at the same tone. However, if you select DIRECT FREQ the VFO frequency will not be offset. The CW tone will be at the zero-beat frequency and will not be heard when you change to SSB. I believe that the default PITCH OFFSET setting is preferable. It means that you can tune in a CW signal while on SSB and then switch to CW without losing the signal.

CW signals are received at an offset equal to the 'pitch' setting of the <PROC> <CW PITCH> control. I set my CW pitch control to 750 Hz.

PC keying sets the control line for 'CW keying' the transmitter from the USB cable. Assuming that you are using RTS for PTT, you should set the PC keying line to DTR (data terminal ready). Set it to DAKY if you are using pin 3 of the RTTY/Data jack for keying.

TIP: If you are sending CW from an external device, the BK-IN button does not have to be turned on for the signal to be transmitted.

TIP: The CW keying signal is sent via the DTR line on the 'Standard' COM port. Not the 'Enhanced' COM port that you are using for CAT.

QSK delay time sets the delay before the transmitter starts to transmit the CW signal. This time is important when you are using a linear amplifier. It must be long enough for the amplifier to switch from receive to transmit before RF power arrives. If it is too short the amplifier will be 'hot switching' which is very bad for the relay contacts. It could also truncate the first dit. If you have an old valve (tube) amplifier it is possible that the default setting of 15 ms will not be enough. You should increase the delay and avoid full break-in operation. If your amplifier is tripping every time you send a CW signal try increasing the QSK DELAY TIME. Note that if the keyer is set at 45 wpm or higher the QSK DELAY TIME will be 15 ms regardless of the setting you have selected.

Note: <FUNC> <BK-IN DELAY> sets the semi break-in delay. That is the time after sending a character before the radio switches back to receiving. While QSK DELAY TIME sets the delay before the transmitter starts to send the CW signal.

CW indicator turns the CW tuning indicator below the S meter on or off. The tuning indicator is handy for tuning a CW signal exactly on frequency. However, you can turn it off if you find it distracting.

CW SETTINGS – MODE CW			
Function	Range	ZL3DW Setting	My Setting
AF TREBLE GAIN	-20 dB to +10 dB	0 (default)	
AF MIDDLE GAIN	-20 dB to +10 dB	0 (default)	
AF BASS GAIN	-20 dB to +10 dB	0 (default)	
AGC FAST DELAY	20-4000 ms	160 ms (default)	
AGC MID DELAY	20-4000 ms	500 ms (default)	
AGC SLOW DELAY	20-4000 ms	1500 ms (default)	
LCUT FREQ	Off / 100 – 1000 Hz	250 Hz (default)	
LCUT SLOPE	6 or 18 dB/octave	18 dB/oct (def)	
HCUT FREQ	Off / 700-4000 Hz	1200 Hz (default)	
HCUT SLOPE	6 or 18 dB/octave	18 dB/oct (def)	
CW OUT LVL	0-100	50 (default)	
CW AUTO MODE	OFF/50M/ON	OFF (default)	
CW BK-IN TYPE	SEMI/FULL	SEMI (default)	
CW WAVE SHAPE	4 / 6 / 8 ms	6 ms (default)	
CW FREQ DISPLAY	DIRECT/PITCH	PITCH OFFSET	
PC KEYING	OFF/DAKY/RTS/DTR	DTR	
QSK DELAY TIME	15/20/25/30 ms	15 ms (default)	
CW INDICATOR	ON/OFF	ON (default)	

KEYER

The Keyer sub-menu sets the parameters of your Morse key or paddle. You can use a 'straight key, a 'Bug,' or a 'Paddle.' There are the usual Iambic choices and an ACS mode which automatically controls the spacing, improving the quality of your Morse. I expect it takes some getting used to.

Keyer type sets the type of Morse key. You can have a different type plugged into the back of the radio.

- OFF is for a straight key. The dots contact (Tip) is used. Or you can just turn <FUNC> <KEYER> off.

- BUG uses the keyer to send the dots but lets you time the dashes.

- ELEKEY-A (Iambic mode A). The keyer finishes sending the last symbol, a dot or dash, and stops. If you were sending a dot and you press both sides of the paddle then let go, the keyer should send an 'A.' If you were sending a dash and press both sides of the paddle then let go, the keyer should send an 'N.'

- ELEKEY-B (Iambic mode B). The keyer finishes sending the last symbol, a dot or dash, and then sends the opposite symbol. If you were sending a dot and you press both sides of the paddle then let go, the keyer should send an 'R.' If you were sending a dash and press both sides of the paddle then let go, the keyer should send a 'K.' Mode B is quite good for sending CQ.

- ELEKEY-Y (Iambic mode Y). Is a variant on mode B. If it is transmitting a dash, the keyer ignores the first dot when you squeeze the paddles.

- ACS makes sure that the character spacing remains correct. It forces a space the length of a dash between characters.

Keyer dot/dash reverses the dots and dashes on a Bug or Paddle.

- NOR (normal) sets the dots to the left paddle and dashes to the right paddle.

- REV (reverse) sets the dashes to the left paddle and dots to the right paddle.

CW weight selects the ratio of the length of a dash compared to a dot. The standard dash is three times as long as a dot, but you can set it anywhere from 2.5 to 4.5 times.

The **Number style** is used to format the contest number in CW message macros. If you don't use the macros in CW contests you can ignore this setting.

- 1290 sends the contest numbers using standard Morse code.
- AUNO sends A instead of one, U instead of two, N for nine, and O for zero.
- AUNT sends A instead of one, U instead of two, N for nine, and T for zero.
- A2NO sends A instead of one, N for nine, and O for zero.
- A2NT sends A instead of one, N for nine, and T for zero.

- 12NO sends N for nine, and O for zero. Other numbers are sent normally.
- 12NT sends N for nine, and T for zero. Other numbers are sent normally.

Contest number: If you are using the CW macros in a contest, set this number to 1 before the contest starts. It will increment as the contest numbers are given out. It is only used to format the contest number in CW message macros. You can decrement the contest number using the DEC key on the FH-2 keypad or CW Keyer.

CW memory 1-5: You can load messages into the CW keyer in two ways, but not from this screen. Why? I don't know. It would have made sense to be able to touch and hold a Soft Key to edit the contents. If the memory slot is set to TEXT, you can type in a message using the on-display, or an external, keyboard. If the memory slot is set to MESSAGE, you can use the CW paddle or Morse key to record a message. See page 17 for setting up the keyer macros.

Repeat interval: If you touch and hold any of the message memories displayed in the MESSAGE dialogue box, the message will repeat at intervals set by this menu item, until you stop the message by sending a dot or dash from the paddle or touching the message memory key again. The delay can be adjusted from 1 second to 60 seconds.

CW SETTINGS – KEYER			
Function	Range	ZL3DW Setting	My Setting
KEYER TYPE	See keyer type above	ELEKEY-A	
KEYER DOT/DASH	NOR/REV	NOR (default)	
CW WEIGHT	2.5-3.5	3.0 (default)	
NUMBER STYLE	See number style	12NT	
CONTEST NUMBER	1-9999	1 (default)	
CW MEMORY 1-5	See CW memory	TEXT	
REPEAT INTERVAL	1-60 seconds	5 sec (default)	

DECODE CW

The internal CW decoder has AFC (automatic frequency control) to help with decoding accuracy. The DECODE CW adjustment sets the bandwidth for the AFC action. The risk in using the wide 250 Hz setting is that in a busy contest the decoder might lock to the wrong CW signal. I plan to stay with the default 100 Hz setting unless I need to change it for some reason. You can choose 25, 50, 100, or 250 Hz.

Operation settings

The Operation Setting menu, <FUNC> <OPERATION SETTING>, has settings relating to QMB, memory groups, split, language, DSP, transmitter power, microphone equalizer, VOX, tuning rate, tuning steps, and audio setup.

TIP: If you make changes to any of the menu settings, it is a good idea to head <FUNC> <EXTENSION SETTING> <SD CARD> <Menu Save DONE> <NEW> <ENT> and save your changes to the SD card.

GENERAL SETTINGS
<FUNC> <OPERATION SETTING> <GENERAL>

NB width changes the duration that the receiver audio will remain attenuated after the noise blanker responds to a noise spike. It can be set to 1, 3, or 10 ms. The short setting can be used for sharp pulse noise such as an electric fence or car ignition noise. The longer setting is better suited to longer nose bursts such as lightning.

NB rejection sets the depth of noise blanker attenuation. You can set 10, 30, or 40 dB of attenuation. The 30 dB default setting should be fine most of the time. If the interference is not too strong a 10 dB setting may be adequate. The DSP noise reduction algorithm looks forward in the receiver audio digital data stream. When the noise level triggers a response, it attenuates the audio for a period set by the NB width control. This introduces a small, unnoticeable, delay (latency) to the signal.

*TIP: the noise blanker **NB Level** is not set in the Operation Setting menu. You can assign it to the FUNC knob for a few seconds by press and holding the NB button. The noise blanker level is the threshold level at which the noise blanker will respond to noise spikes.*

Most DSP noise blankers (NB) work by eliminating or attenuating noise peaks that are above the average received signal level. They usually have no effect on noise pulses that are below the average speech level. The noise blanker is best used to reduce or eliminate regular pulse-type noise such as car ignition noise. You can experiment with the NB rejection level and the NB width controls when tackling a particular noise problem. If the noise blanker has no noticeable effect, try the DNR (digital noise reduction) button.

Beep level sets how loud the beeps are. Set (0-100). You can make them loud! The default is 10.

RF/SQL VR sets the function of the RF/SQL control. The outer knob controls either the RF gain of the receiver or the squelch.

The default setting is for it to act as an RF Gain control. As an RF Gain control, it provides manual adjustment of the gain of the RF and I.F. stages. It can be very useful to turn down the RF gain to improve the signal to noise ratio of reasonably strong signals, especially on the noisy lower frequency bands. But most people leave it turned up to maximum all the time. In fact, the Yaesu manual states that the *"RF/SQL knob is normally left in the fully clockwise position."* I find it much more useful to change the menu setting so that the control operates the receiver squelch. If you operate mostly on 40m, 80m, and 160m I recommend setting the menu to the **RF** gain position. If like me you favor the higher bands, set the menu to **SQL**.

Tuner select is used to tell the radio that you have an external Yaesu compatible antenna tuner or an ATAS 120A active tuned antenna, connected.

INT indicates the internal antenna tuner is available. EXT means that an external Yaesu antenna tuner is connected. And you should select ATAS if you are using the ATAS 120A active tuned antenna. See using the tuner on page 93.

The ATAS tuner receives tuning information over the antenna cable, so it won't tune unless you set this menu option to ATAS. The Yaesu FC-40 antenna tuner receiver receives tuning information over the supplied control cable connected to the TUNER jack on the radio.

TIP: if you are using a non-Yaesu antenna tuner, you should set TUNER SELECT to INT. Although, if the Yaesu tuner or ATAS antenna is not connected, the radio reverts to using the internal antenna tuner anyway.

The **232C rate** and **232C time out timer** settings relate to the serial port settings for the 9 pin DB-9 serial port on the back of the radio. They do not affect the USB connection. The 232C rate sets the baud rate in bits per second and the 232C time out timer sets the time that the radio will wait for a response to an RS-232 command.

CAT settings: CAT RATE is used to set the interface speed of the USB connection between the radio and the PC software. The speed should match the speed specified in the software application. It does not need to be very fast. 9600 bps (bits per second or 'bauds') is adequate for most connections. The default is 38400 bps. CAT TIME OUT TIMER sets the time that the radio will wait for a response to a CAT command. Leave it at the default setting unless you are experiencing CAT control problems.

Neither WSJT-X nor MixW will control the radio unless CAT RTS is turned OFF. The Preset mode turns it off by default. CAT RTS is not the PTT control line. That is set elsewhere and relates to the second COM port. This one is used as an interrupt.

When it is set to 'ON' the radio monitors the RTS line on the CAT COM port and will respond to changes initiated by the PC software. When it is set to 'off,' the radio does not monitor the RTS line on the CAT COM port.

QMB CH: sets the number of Quick Memory Bank slots associated with the QMB button. The default is 5 channels, but you can increase it to 10 channels. Each new entry is saved into the top memory slot. All the previously saved frequencies move down one place and the one at the bottom of the list is discarded.

You can use this feature as a 'Scratch pad' to save a frequency that you want to return to. For instance, a station that you want to work but which is in a QSO with somebody else. Or you can use it to store frequencies that you use a lot, such as Net frequencies.

Turning on **MEM GROUP** splits the 99 stored memory slots into five groups. When they are split, you can press V/M to change to the memory channel mode, then select <FUNC> <M-GROUP> to use the FUNC knob to choose a memory group. Then you can scan a specific group or step through the memory slots in the chosen group using the microphone UP and DWN buttons. Press and hold to start a scan.

You can also press and hold the STEP/MCH button and step through the channels in the selected group using the VFO tuning ring.

It is quite handy to save all the stored frequencies for a particular band into a group. That way the radio will not jump all over the HF band when you scan the memory slots. For example, you might put all the 6m repeater and simplex channels into group five. Or you might put all the international beacon frequencies into group two.

There are five general-purpose groups and two special groups. You cannot save a frequency to a memory position and then select which group it goes into. Instead, you must save the frequency into a memory slot that is in the wanted group. If you want to put a frequency into group 5 you have to store it in one of the slots between 80 and 99. You can't name, the groups either.

Memory channel	Group
1-19	1
20-39	2
40-59	3
60-69	4
80-99	5
Nine PMS# scan edges	(PMS)
Ten* 5 MHz (60m band) channels	5M

PMS = programmable memory scan. The eighteen memory slots hold the lower and upper frequency limits for nine scan ranges.

* The US version has ten 5 MHz (60m band) channels. The UK version has seven 5 MHz channels. Other variants may not have any 5 MHz channels at all.

NOTE: You cannot use the <FUNC> <M-GROUP> menu item to change between groups unless the radio is in memory channel mode. In the VFO mode, you can display but not change the current group setting. This is mildly annoying and to my mind an unnecessary constraint.

If you have Groups turned on, the microphone UP and DWN buttons can only select memory slots that are within the currently selected group. To select a memory channel in a different group, you have to select the 'memory channel mode' with V/M, then select the group with M-GROUP, and then select the stored channel. Could it be more difficult?

Split settings: The three split settings affect what happens when you press and hold the SPLIT button. I found the descriptions in the manual rather confusing, and it took a while to work it out.

WARNING: It is very easy to transmit on the wrong band or mode. If you press the SPLIT button, the transmitter will shift to the alternate VFO, usually VFO-B, but it will not change the frequency or mode of that VFO. The transceiver will happily transmit on whatever frequency and mode VFO-B is on. Always press and hold the SPLIT button to set the split offset.

QUICK SPLIT INPUT acts differently from what I expected. ON means quick split is <u>not</u> enabled and OFF means that quick split is enabled.

If you press and hold the SPLIT button when QUICK SPLIT INPUT is ON, a popup box appears on the screen allowing you to select the split offset that you want. Touch 5 and then kHz if you want a five-kilohertz split. Touch -, 2, and then kHz if you want a negative two-kilohertz split.

If you press and hold the SPLIT button when QUICK SPLIT INPUT is OFF, VFO-B will be set to transmit at the offset determined by the QUICK SPLIT FREQUENCY. Press and holding the SPLIT button a second time moves the transmitter further away from the receiver frequency by the same amount. For example, if the offset is 5 kHz, the transmitter frequency will move up 5 kHz each time you press and hold the SPLIT button. I am not sure that is particularly useful since you can easily adjust the transmitter VFO after using the A/B button or while holding down the TXW button. I plan to leave this setting in the default OFF position.

QUICK SPLIT FREQ sets the 'quick split.' It is only relevant if QUICK SPLIT INPUT is set to OFF. In other words, if 'quick split' is enabled. You can set it anywhere from -20 kHz to +20 kHz. The default is 5 kHz which is good for upper sideband operation on SSB. You would probably set a 2 kHz offset for CW operation and a negative split for lower sideband. You would reverse the split if you happen to be the DX station working a pile-up.

The **TX time out timer** limits the time that the transmitter will transmit continuously. You can set it to OFF, or anywhere between one minute and thirty minutes. If you are prone to long overs, this might be a benefit. You get a beep warning when you are close to the set time and the transceiver will return to receiving after the time has elapsed. Releasing the mic PTT every few minutes lets the 'other bloke' have a go and resets the timer.

It can be a good thing to use the timer if you are controlling the transmitter from a digital modes program or a connected device. Sometimes software or hardware glitches may cause an application to hang, and the transmitter will go into a permanent transmit mode. In the SSB Data mode, there probably won't be any modulating signal, but if it is a CW, RTTY, or FM signal the radio could be transmitting at full power.

TIP: it is essential to set the TX time out timer if the radio is mounted in a vehicle. A microphone slipping down the back of a car seat or getting dragged sideways by the mic cord so that the PTT switch is operated is a classic way of accidentally transmitting for hours on end. Broadcasting the car radio and your conversation with the dog.

Microphone scanning functions: The supplied SSM-75E hand microphone and many after-market microphones have handy buttons for controlling functions on the radio. The UP and DWN buttons on the top of the hand microphone step the VFO up or down by 10 Hz, (or 5 Hz). Or to the next memory channel if the radio is in memory channel mode. Press and hold either UP or DWN to start a scan.

MIC scan resume affects what happens when the receiver encounters a signal during a scan that has been initiated by press and holding the UP or DWN button on the microphone.

Mode	MIC SCAN RESUME	A signal is received, or the squelch is open	No signal received and squelch closed
SSB or CW	PAUSE	Slower scan	Fast scan
SSB or CW	TIME	Slower scan	Fast scan
FM and other modes	PAUSE	The scan stops until the signal disappears and the squelch closes	Fast scan
FM and other modes	TIME[1]	Scan waits for 5 seconds then starts to scan. If the squelch is still open it will pause again. If the squelch is closed it will resume scanning.	Fast scan

Note 1: as far as I can tell you can't change the 5 second pause time.

TIP: turn the VFO a ¼ turn while scanning to reverse the scan direction. Anticlockwise makes the scan go down and clockwise makes the scan go up.

The **MIC scan** setting affects the press and hold function of the UP and DWN buttons. With the control set to ON, scanning will continue until you press UP or DWN again, or press the PTT switch, or touch the spectrum display. If you press and hold UP or DWN with the MIC scan control set to OFF, the VFO will tune up or down rapidly until you release the button. But it does not initiate a scan.

Ref frequency fine adjustment: The FTDX10 does not have an input for the connection of an external reference frequency oscillator. It doesn't need one because the local oscillator is a 0.5 ppm high stability TCXO (temperature-controlled crystal oscillator). The TCXO in the radio is much more accurate than many frequency counters. I recommend that you do not make any changes to this setting unless you are 100% confident that fiddling with it is not going to make the frequency accuracy of your transceiver worse.

However, if you feel the need, you can calibrate the receiver frequency by netting to a standard frequency broadcast such as WWV or WWVH and adjusting the REF FREQ FINE ADJ control, (page 180). Or you can measure the transmitter frequency using a frequency counter connected via a suitable coupler and RF power attenuators. The frequency counter must be very stable and/or locked to a GPS referenced clock or a Rubidium frequency standard. Remember to apply any offset that might apply to the transceiver mode. Any adjustment made using FREQ FINE ADJ will affect both the receiver calibration and the transmitter frequency.

Keyboard language: There is a range of different keyboard layout options. The radio defaults to different options depending on the transceiver 'country' version. My radio defaults to ENGLISH (US). As far as I can tell this only affects the keyboard mapping of external keyboards plugged into one of the USB ports. The internal keyboard, displayed when setting up text macros, is unaffected by this setting.

RX DSP

The RX DSP menu provides settings for the filters implemented in the 24 kHz I.F. DSP stage of the receiver.

APF width sets the bandwidth of the CW APF (audio peaking filter). Narrow, Medium, or Wide. A narrow setting will eliminate all other signals, but the signal needs to be precisely tuned. You can fine-tune the filter using the APF knob, or fine tune the VFO using the CW tuning indicator underneath the S meter and the ZIN button. The wider settings are easier to tune but you may hear more noise or interference from other stations operating very close to the frequency.

Contour level: The contour control provides a small dip or peak in the IF filter bandwidth. It is represented with an orange dip (or peak) on the filter function spectrum display. You can use the CONT knob to move the dip (or peak) to any frequency across the IF bandwidth. The function can be used as a tone control to boost or reduce the treble or bass response at the high or low end of the IF passband. Or it can be used as a 'mini notch filter' to reduce weak interfering signals.

The control is adjustable from a -40 dB dip to a +20 dB boost, with a default of -15 dB. *This range is not correctly stated in the current issue of the Yaesu manual.*

Contour width sets the width of the contour dip or peak. I can't see much point in changing from the Q = 10 default setting. But you can adjust it from 1 to 11, "just like in the Spinal Tap movie." I guess a narrow setting might be useful if you are using the Contour control to reduce low-level interference.

I.F. notch width is used to change the IF notch from the default wide setting to a narrow setting. The wide setting is great for noisy or multi-carrier interference such as ADSL line noise. The narrow notch is better for single carrier "birdies," and it will have less effect on the quality of the wanted signal. I set the notch width to 'narrow' and only change it to 'wide' if the notch is not eliminating the interfering signal effectively.

TX AUDIO

The **AMC release time** sets the decay time after the AMC acts to reduce excessive microphone audio levels. During the SSB audio setup, it is easier to adjust the AMC LEVEL control with the COMP meter if the release time is set to FAST. The default level is MID. When operating, I can't tell the difference.

The **Parametric Equalizer** settings were discussed on page 14. You can tailor the audio frequency response of your transmitted signal using the three-band parametric equalizer. There are settings for when you have the speech compressor turned off and for when you have it turned on.

TX GENERAL

HF, 50M, and 70M max power sets the maximum power of the transmitter on the relative bands. You might want to reduce power to meet your license conditions or to ensure that you don't overdrive your linear amplifier.

TIP: turning down the Max Power leaves a dead zone at the top of the <FUNC> <RF POWER> control. If you reduce the RF POWER control to the same output power level as the Max Power control and then reset Max Power to 100W, the transmitter power will stay low. However, if you don't adjust the RF POWER control, and leave it set to its maximum when you reset Max Power to 100W, the transmitter power will return to full power.

AM max power does what it says. It is adjustable from 5 to 25 watts.

VOX select is normally set to operate on input from the microphone (MIC), but you can set it to DATA so that it triggers on an audio signal applied to the USB cable or the RTTY/DATA jack. You could use the latter mode if you are unable to get the RTS/DTR signaling to work with your particular digital mode software. Or if you wanted to transmit a recording.

When VOX SELECT is set to DATA, the **DATA VOX GAIN** control sets the sensitivity of the VOX when a data signal is received. It should be set so that the transmitter operates when a digital signal is sent from the PC software but not when no signal is being sent. I expect that the default level of 50 will be fine, 99% of the time.

Emergency Freq TX (ON or OFF) allows the radio to transmit on the Alaska emergency frequency of 5167.5 kHz. This is disabled on some transceiver versions.

TUNING

This group sets the tuning step rate and the optical encoder setting for the VFO knob and the MPVD "VFO tuning ring."

- SSB/CW DIAL STEP sets the VFO to 10 Hz or 5 Hz per tuning step and the VFO tuning ring to 100 Hz or 50 Hz per step.

- RTTY/PSK DIAL STEP sets the VFO to 10 Hz or 5 Hz per tuning step and the VFO tuning ring to 100 Hz or 50 Hz per step.

- CH STEP sets the STEP/MCH channel tuning step for the VFO ring to, 1 kHz, 2.5 kHz, 5 kHz, or 10 kHz. I selected 1 kHz.

- AM CH STEP sets the AM channel tuning step for the VFO ring to, 2.5 kHz, 5 kHz. 9 kHz, 10 kHz, 12.5 kHz, or 25 kHz.

- FM CH STEP sets the FM channel tuning step for the VFO ring to, 5 kHz. 6.25 kHz, 10 kHz, 12.5 kHz, 20 kHz, 25 kHz. I selected 25 kHz.

- MAIN STEPS PER REV sets the VFO steps per revolution, 250, 500, or 1000.

- MPVD STEPS PER REV sets the VFO tuning ring steps per revolution to, 250, or 500.

The FINE button changes the VFO tuning rate to 1 Hz. It does not affect the tuning rate of the VFO ring. The STEP/MCH button is ignored if FINE is turned on.

The STEP/MCH button turns on the STEP mode and is indicated on the LED strip. The VFO tuning rate is unaffected, but the VFO tuning ring changes to the CH STEP tuning rate. Press and hold the STEP/MCH button in VFO mode to change the VFO tuning ring tuning rate to 1 MHz per step. This tuning rate is indicated with a flashing orange LED on the STEP/MCH button.

Display settings

The Display Setting tab, <FUNC> <DISPLAY>, includes settings for the display, scope, and external monitor. These include display brightness, screen saver, mouse speed, the RBW, bandwidth, and sensitivity of the spectrum and waterfall display, and the resolution setting for an external monitor.

TIP: If you make any changes to the menu settings, it is a good idea to head to <FUNC> <EXTENSION SETTING> <SD CARD> <Menu Save DONE> <NEW> <ENT> and save a backup to the SD card.

DISPLAY

My call lets you enter a 12 character personal message which is displayed on the 'opening screen' during the transceiver start-up. I had just enough room to enter ZL3DW FTDX10. **My call time** sets the time that your message will be displayed. (OFF, 1, 2, 3, 4, or 5 seconds). Naturally, setting a longer time lengthens the time it takes for the transceiver to start. The actual time that the information is shown seems to be a little longer than the time you specify.

Screen saver sets the time before the screen saver starts. The timer resets if you touch the screen or make any change that affects the screen such as changing bands or pressing a button. The default is 60 minutes, but you can set the delay to OFF, 15, 30, or 60 minutes. I don't recommend turning the screen saver off because you could eventually get a screen image burnt onto the display.

LED dimmer sets the brightness of the LEDs, from 0 to 20. This is a nice feature. You may want to dim the LEDs if you are working at night, especially if it's a contest and your eyes are tired. I set it to 8, a little lower than the default level of 10.

Adjusting the screen brightness & contrast: Pressing <FUNC> <DIMMER> and using the FUNC knob sets the brightness of the touchscreen between 0 and 20. (The default is 10). Pressing <FUNC> <CONTRAST> and using the FUNC knob sets the contrast of the touchscreen between 0 and 20. (The default is 10).

Mouse pointer speed does what it says. It adjusts the sensitivity of the mouse from 0 to 20 with a default setting of 10. I don't bother using a mouse because it is so limited, so I have not changed the setting.

SCOPE

RBW (Resolution bandwidth): There are three choices of resolution bandwidth for the spectrum scope. I can't see any reason why you would want to change from the default HIGH setting. <FUNC> <DISPLAY SETTING> <SCOPE> <RBW>.

The **Scope center** control changes the way that signals are displayed on the spectrum and waterfall display. *TIP: the FILTER setting is horrible. Don't use it!*

I strongly recommend using the CARRIER 'carrier point' option which places the real VFO frequency on the center line or marker and displays the receiver passband on the right for USB, on the left for LSB, or in the center for AM, CW, or FM.

If you choose the FILTER option and the CENTER display, the receiver passband will be placed in the center of the screen. If you turn on <FUNC> <MARKER>, the marker indicating the carrier point will be displayed on the left side of the receiver passband for USB, or the right if the radio is set to LSB. If you choose the FILTER option and the CURSOR or FIX display, the marker indicating the carrier point will be displayed in the center of the receiver passband.

To change the scope center point, select <FUNC> <DISPLAY SETTING> <SCOPE> <SCOPE CTR> and select (CAR POINT or FILTER).

2D and 3DSS sensitivity: There are 'HI' and 'NORMAL' sensitivity settings for the 3DSS or waterfall and spectrum display. Try both settings and see which you prefer.

<FUNC> <DISPLAY SETTING> <SCOPE> <2D DISP SENSITIVITY>

<FUNC> <DISPLAY SETTING> <SCOPE> <3DSS DISP SENSITIVITY>

I'm not sure why Yaesu went with HI when they could have used HIGH. However, I prefer the HI setting for both the 3DSS display and the 2D display.

EXT MONITOR

Ext display: You must change a menu setting to use an external monitor.

<FUNC> <DISPLAY SETTING> <EXT MONITOR> <EXT DISPLAY> ON

Pixel: I recommend you select the higher screen resolution option. My monitor does not have an 800x480 option, so it just displays everything bigger.

<FUNC> <DISPLAY SETTING> <EXT MONITOR> <PIXEL> 800x600 or 800x480

800x600 pixels looks pretty chunky on a big monitor. You get a big, low-resolution, image that looks the same as the touchscreen. It does not have touchscreen capability and you cannot change the transceiver controls like you can with SDR software. It might be useful if you are demonstrating the radio, or you have poor eyesight… but I don't see any appeal in trading the high-resolution touchscreen for a big blocky image on a monitor.

Extension settings

The Extension Setting menu, <FUNC> <EXTENSION SETTING>, sets the date and time, allows you to save and recall menu and memory contents from the SD card, list, update the current firmware, enter the calibration mode, and if all else fails, engage one of the three levels of reset.

DATE & TIME

Setting the time and date. The radio has a clock, but it is never displayed on the screen. It is only used for time-stamping saved files.

Set the DAY, MONTH, and YEAR and **press <FUNC> to save the date**.

Then set the HOUR, and MINUTE and **press <FUNC> to save the time**.

If you don't press FUNC both times, the date and/or time will slip.

If you want files date stamped with UTC date and time, enter UTC details. I believe it is easier to use local time for filenames.

SD CARD

The **SD Card** settings have already been discussed in the 'Using an SD card' chapter, on page 59 and the 'Firmware Update' chapter on page 61.

TIP: if you insert the SD card with the transceiver turned on, you will be asked if you want to 'SETUP?' It is safe to answer YES or NO. If you touch YES, the SD Card setup menu will open.

SOFTWARE VERSION

The SOFT VERSION option displays the currently installed firmware revisions.

Your firmware should be,

- Main CPU   V01-08 or newer
- Display CPU   V01-03 or newer
- DSP   V01-01 or newer
- SDR (FPGA)   V02-00 or newer
- AF DSP   V01-00 or newer

CALIBRATION

The calibration section is for resetting the touchscreen calibration. You should only do this if a section of the touchscreen has stopped working correctly or part of the screen is not being displayed. For example, if the bottom set of Soft Key controls are partly hidden. Touchscreen calibration is very easy to perform.

Select <FUNC> <EXTENSION SETTING> <Calibration DONE>

A + sign will appear at the top left of the display. Touch the + symbol and it will reappear at the top right. Touch that and it will move to the bottom left, then the bottom right, and finally the center of the display. Touch each in turn and the display will be calibrated.

RESET

There are three levels of software reset available and a hardware reset. You should only do a hardware reset as a last resort. Check the radio settings and connected cables first. Freeze-ups are usually fixed by turning off the radio and then restarting it. If the radio will not turn off, you should turn off or disconnect the DC supply. Then reconnect the DC and restart the radio.

If that fails, disconnect the CW key or paddle, all control cables, and the USB cable, leaving only the antenna cable and the power supply cable connected. Turn off the transceiver. Turn off the DC supply or unplug the DC cable. Wait for 30 seconds, then reconnect or turn on the DC supply and turn on the radio. If you still have problems. Check the tightness of the DC cables where they are connected to the DC supply. Several readers have experienced problems with poorly crimped DC lugs on the power supply end of the cable, (not the radio end of the cable).

Push <FUNC> <EXTENSION SETTING> <RESET> (reset option) <DONE>

Memory clear resets all of the stored memory slots. Memory channel M-01 will be reset to 7.000 MHz LSB. The ten, or seven, 5 MHz channels (60m band) will be unaffected. You can restore the memory channels from a saved backup on the SD card. See the 'Using an SD card' chapter, on page 59.

Menu clear resets all menu settings to the factory defaults. You can restore the menu commands from a saved backup on the SD card. See the 'Using an SD card' chapter, on page 59. *This command does not affect the saved memory channels.*

All reset clears the radio back to default settings including the menu settings and the memory channels. This reset will cause the radio to shut down and restart.

Hardware reset. If all else fails, turn the transceiver off, then turn it on again while holding down the FINE and LOCK buttons. When the transceiver starts, release the two buttons.

Front panel controls

This chapter describes the front panel controls. I have provided a few additional details to supplement the information provided in the Yaesu manuals. The front panel controls are placed in functional groups making it easy to find what you want.

POWER

I guess you have already worked this one out. If you have entered your callsign it will be displayed under the Yaesu logo on the start-up splash screen. (See page 8).

Press and hold the POWER button to turn the radio on or off.

TUNE

Pressing TUNE activates the automatic antenna tuner. A red TUNE icon is displayed below the upper Soft Keys on the touchscreen. The antenna tuner will not automatically initiate an antenna matching (tuning) operation if the SWR is high. You have to press and hold the TUNE button to make the tuner 'tune.' It will attempt to match the complex impedance of the antenna plus the feeder cable to the 50 ohm impedance of the transmitter.

Make sure that your antenna is connected, and the frequency is clear. A carrier of about 10 Watts is generated and the red 'TUNE' icon flashes while the tuner is operating. You will hear relays clicking as the tuner attempts to find a match.

The tuner remembers the settings for each 10 kHz band segment that you transmit in. There is more about using the tuner on pages 75 and 93.

When a Yaesu FC-40 antenna tuner or an ATAS-120A active tune antenna is connected, the TUNE button will control that instead of the internal tuner.

VOX/MOX

VOX stands for voice-operated switch. When VOX is on, the radio will transmit when you talk into the microphone without you having to press the PTT button. Using VOX is popular if you are using a headset or a desk microphone.

VOX can be activated when the radio is set to SSB, AM, FM, or DATA. Pressing the VOX/MOX button turns the VOX on or off. An amber indicator on the button indicates VOX mode. VOX PTT switching can also be used by external digital mode software if the RTS/DTR keying lines cannot be configured successfully. The full set of VOX settings is covered on page 76.

Press and holding the VOX/MOX button enables the MOX function. MOX stands for 'manually operated switch.' The switch is effectively wired in parallel with the microphone PTT switch. After engaging MOX, the radio will keep transmitting until you turn it off again. The orange LED on the button does not light when MOX is active, but the radio will be transmitting and that is indicated by the red TX icon on the LED strip.

In CW mode MOX will allow you to send Morse code from a CW message, or the paddle or Morse key without having BK-IN turned on.

If you turn MOX on in the voice modes, the microphone will be live, so you will transmit any conversations, phone calls, talking to the cat, praising the dog, background music, or audio from another radio. The voice keyer will operate while the radio is being keyed via MOX, but as usual, BK-IN must be turned on.

FUNC

Pressing the FUNC knob opens the Function menu, which is described in the Function Menu chapter and the five following chapters, starting on page 113.

Pressing the FUNC knob when a menu is already being displayed, selects the highlighted option. It may toggle a value such as KEYER or BK-IN, or it can allocate an adjustment such as PROC LEVEL to the FUNC knob. Turning the FUNC knob adjusts the level selected function. It usually opens a pop-up window or a sub-menu like the COLOR options. The currently allocated function is displayed in blue text under the VFO numbers at the top-right of the display.

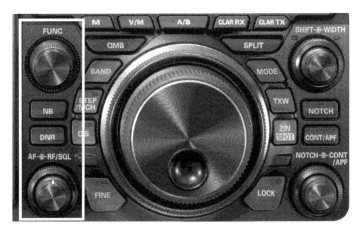

Figure 42: FUNC, noise reduction, and audio controls

Other items may pop up for a few seconds. For example, press and holding the NB button pops up a level setting option that can be changed by turning the FUNC knob. After a few seconds, the popup disappears, and the selectable function returns to the previously allocated function.

Most popup windows only stay active for four seconds, so you have to be quick to start adjusting before the option disappears.

NB (NOISE BLANKER)

Pressing the NB button turns on the noise blanker. It is indicated with an orange LED on the button. Holding down the NB button down brings up a sub-menu where you can adjust the NB level, which is really the noise blanker threshold. The noise blanker is designed to reduce or eliminate regular pulse-type noise such as car ignition noise. You may need to experiment with the NB level and the two related menu controls, NB width and NB Rejection, when tackling any particular noise problem. A full description of the noise blanker can be found on page 72.

DNR (DIGITAL NOISE REDUCTION)

The digital noise reduction system in the FTDX10 is very effective. Digital noise reduction is indicated with an orange LED on the button. Holding down the DNR button down brings up a sub-menu where you can adjust the noise reduction level from 1 to 15. The Yaesu manual states that there are "15 different noise-reduction algorithms" which suggests that the control is not just a linear progression and that some of the intermediate settings may work better than the high or low options. See page 73 for more details.

AF AND RF/SQL

The black inner knob controls the AF Gain (volume) and the outer RF/SQL knob controls either the RF Gain or the Squelch.

➢ **RF gain and squelch**

The default setting for the RF/SQL control is for it to act as an RF Gain control. The RF Gain control provides a manual adjustment of the gain of the RF and I.F. stages.

It can be useful to turn down the RF gain to improve the signal to noise ratio of reasonably strong signals, especially on the noisy lower frequency bands. But most people leave it turned up to maximum all the time. In fact, the Yaesu manual states that the "RF/SQ knob is normally left in the fully clockwise position." I find it much more useful to change the menu setting so that the knob operates as a receiver squelch control.

My recommendation is to choose the RF gain option if you operate mostly on 40m, 60m, 80m, and 160m bands. If like me you favor the higher bands, set the menu to SQL. <FUNC> <OPERATION SETTING> <GENERAL> <RF/SQL VR> (RF or SQL).

There is no indication of the squelch level when SQL is selected, but a green BUSY indicator is displayed on the LED strip when the squelch is open.

MEMORY AND CLARIFIER CONTROLS

Figure 43: Memory and Clarifier controls

M

The M button is used to save the current VFO settings to a memory channel. You can also use it to select a previously saved channel. For a detailed description see the 'Memory groups and channels' chapter on page 98.

➢ **To save a frequency to a memory slot.**

Set the mode and tune the VFO to the frequency that you want to save. Press M to bring up the memory channel list. Use the FUNC knob to find a blank memory slot or a slot you want to overwrite. Press and hold the M button to write the frequency into the selected position. Add a name and touch ENT or change the scan setting. Touch BACK to exit.

➢ **To recall a memory channel**

To recall a memory channel, press M to bring up the memory channel list. Use the FUNC knob to find the wanted channel, then press the FUNC knob to change the radio to the memory channel mode on the selected memory channel.

V/M

V/M swaps between the VFO mode and the memory channel mode.

While in the memory channel mode, press and hold the STEP/MCH button to step through the memory channels by tuning the VFO tuning ring.

This mode is indicated with a flashing orange LED on the STEP/MCH button. If Groups are turned on, you can only step through channels in the selected group.

Alternatively, while in the memory channel mode, you can step through the memory channels using the microphone UP and DWN buttons or start a memory channel scan by holding down one of those two buttons.

If you want to see a list of the memory channels instead of stepping through them, use the M button instead.

A/B

The A/B button usually toggles between VFO-A and VFO-B. You can leave one VFO set to a frequency that you want to return to while tuning around with the other one. For example, you might leave VFO-A tuned to a DX station and use VFO-B to listen to the calling stations or to check activity on a different band during a contest.

Press and holding the A/B button sets both VFOS to the currently displayed frequency and mode.

Note that I said the A/B button "usually" toggles between VFO-A and VFO-B. It actually just swaps the VFOs, so if either VFO is in the memory channel mode, pressing A/B toggles between the MEM mode and the VFO. If both VFOs are in the memory channel mode, pressing A/B toggles between the two memory channels. I guess that could be handy if you want to toggle the radio between two memory channels.

TIP: during SPLIT operation, the FTDX10 uses VFO-A for receiving and the VFO-B for transmitting (or vice versa). While using VFO-A, set the receive frequency so that you can hear the DX station. Press and hold the A/B button to sync the two VFOs, then press A/B and tune the VFO knob to set your transmitter frequency on VFO-B. You will be able to hear stations in the pileup. When you have selected a suitable transmit frequency press A/B once again to return the receiver to VFO-A. Finally press the SPLIT button, so that the radio will transmit on the VFO-B frequency.

CLAR RX

The receiver clarifier is equivalent to RIT in earlier radios. It allows you to offset the receiver frequency while leaving the transmitter frequency the same. As you adjust the offset using the VFO tuning ring, the revised receiver frequency is indicated on the VFO display.

The receiver clarifier is useful if you are chatting with a station that is slightly off frequency. Rather than move the VFO which will cause the other station to adjust its frequency the next time you transmit, you use the receiver clarifier to fine-tune the incoming signal. It is especially useful in a net where only one station is a little off frequency. The function is fully described back on page 69.

Press and hold CLAR RX to reset the clarifier offset to zero.

CLAR TX

The transmitter clarifier is equivalent to XIT in earlier radios. It allows you to offset the transmitter frequency from the receiver frequency.

It can be used as a way of operating Split without using the split button. If you are using the CENTER display with the markers turned on, you will see the red transmit frequency marker move away from the center as the offset is increased. In the FIX and CURSOR modes, you will see the red transmit frequency marker move away from the green receiver frequency marker as the offset is increased.

THE VFO CLUSTER

The VFO knob is surrounded by a cluster of buttons. Some of them set the function performed by the VFO tuning ring (MPVD). They are,

- BAND (easier to press BAND and select a band by touching the screen)
- MODE (easier to press MODE and select a band by touching the screen)
- SPLIT (The VFO ring adjusts the transmitter VFO frequency)
- STEP/MCH (VFO ring adjusts the VFO or memory channel in step mode)
- CLAR RX & CLAR TX (adjust the receiver or transmitter clarifier offset), or
- CS (VFO ring adjusts one of the 16 custom selections).

VFO KNOB

This is the 42 mm main tuning knob. You know what it does. The VFO control doesn't feel weighted and there is no apparent 'flywheel' action. The knob does not try to keep spinning when you stop turning the knob. There is a drag adjustment in a slot under the knob if you want to firm up the tuning. But on my radio, the VFO is quite stiff even at the lowest drag setting. Unlike some other radios, there is no acceleration of the tuning rate if you turn the VFO quickly. In fact, the tuning rate seems to max out if you turn the knob very fast. The knob has a small dimple to aid tuning.

MPVD (MULTI-PURPOSE VFO OUTER DIAL)

I call it the 'VFO tuning ring' because it is a better description. It is the outer dial around the VFO knob. The VFO tuning ring can be configured to adjust 23 different functions. Pressing the BAND, MODE, SPLIT, STEP/MCH, CLAR RX, or CLAR TX allocates those functions to the tuning ring. Press and holding STEP/MCH allocates the memory channel step in the memory channel mode or 1 MHz step in VFO mode. Press and holding CS allows you to set one of 16 custom selections.

If none of those options is enabled, the VFO tuning ring tunes the current VFO at ten times the selected VFO tuning rate, i.e. 100 Hz or 50 Hz for SSB, CW, or DATA.

The VFO tuning ring is not affected by the FINE button, but the tuning rate increases to the adjustable 'CH STEP' rate (1 kHz – 10 kHz) if STEP/MCH is pressed. Press and holding the STEP/MCH button sets the tuning rate to 1 MHz steps.

Figure 44: VFO tuning ring and buttons

FINE

The FINE tuning button is located at the lower left of the VFO button cluster. It selects a VFO tuning rate of 1 Hz per step for the SSB, CW, and data modes and 10 Hz per step for AM, FM, and the FM data modes.

CS

CS stands for 'Custom Select.' Press and hold allows you to allocate one of sixteen different functions to the big tuning ring. I usually choose spectrum scope LEVEL. Once allocated, turning on the CS button activates that function. If you use the Clarifier, Split, or STEP/MCH buttons, CS will turn off and the selected function will be allocated to the tuning ring. Pressing CS will reactivate the custom selection.

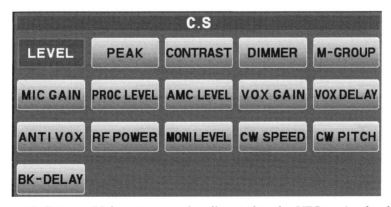

Figure 45: Sixteen CS functions can be allocated to the VFO tuning knob

STEP/MCH

Press the STEP/MCH button to enter the channel step mode, indicted with an orange LED on the button and a white STEP indicator on the LED strip. The channel step mode changes the tuning step of the VFO tuning ring. There are different channel steps for SSB, AM, and FM. See Tuning Steps on page 146.

Press and holding the STEP/MCH button in the VFO mode changes the VFO tuning ring tuning rate to 1 MHz. In the memory channel mode, it lets you step through the memory channels. If Groups are turned on, you can only step through channels in the selected group. This mode is indicated with a flashing orange LED on the STEP/MCH button.

BAND

The BAND button opens a popup menu so that you can change bands. You have to be quick because the popup only stays visible for about four seconds.

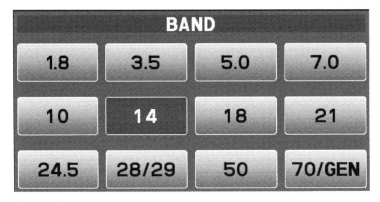

Figure 46: Band selection

Each band has Yaesu's standard 'band stacking register. Selecting a band sets the radio to the last frequency and mode you used on the selected band. Pressing BAND and selecting the same band again moves you to the frequency and mode you used before that, and pressing it again moves you to the frequency and mode you used before that. Each band button holds three frequencies.

TIP: you can manually program the band stacking register. Select a band and tune the VFO to the frequency you want to save. Then press the BAND button and touch the same band icon again to load the frequency into the band stacking register. If desired, you can move to a second frequency, press the BAND button and the same band icon again to load that frequency into the band stacking register. Move to a third frequency and repeat the process to load a third frequency into the band stacking register. It seems like more trouble than it's worth, to me. Note that these memory slots are volatile. The band stack can be overwritten when you switch bands.

You cannot use the band Soft Keys to directly enter a frequency. You can type in a frequency after touching the three Hz digits on the VFO frequency display.

The band edges are preset according to the region that the radio is for. Some models do not include the 5 MHz (60m band) or 70 MHz (4m band) frequencies. See 'Troubleshooting – Enable the 60m and 4m band frequency allocation.' As far as I know, you cannot change the band edges inside a band.

QMB (QUICK MEMORY BANK)

You can use the Quick Memory Bank as a 'scratch pad' to save a frequency that you want to return to. For instance, a station that you want to work but which is currently in a QSO with somebody else. Or you can use it to store a list of frequencies that you use a lot, such as Net frequencies.

Press the QMB button to recall the last saved channel. Subsequent presses step through the five (or ten) saved frequencies. Blank channels are not included. The current QMB memory slot is displayed above the mode indicator. The final press will return the radio to the VFO mode on the previous frequency.

Press and hold the QMB button to save the current VFO settings to the top of the stack. Each time a channel is saved it is placed into slot 1 and all the previously saved frequencies move down one slot. The one at the bottom of the list is discarded.

<FUNC> <OPERATION SETTING> <QMB CH> sets the number of quick memory slots associated with the QMB button. The default is 5 channels, but you can increase it to 10 channels. *Warning: it takes a lot longer to step through ten channels.*

SPLIT

Working in the 'Split' mode is a very common requirement if you are trying to work a rare DX station or DXpedition when they have a 'pileup' of stations calling them.

To work 'Split' you usually set VFO-A to receive the frequency that the DX station is using, and transmit on the frequency indicated by VFO-B. On the bands above 10 MHz, you transmit USB on a frequency that is a few kHz higher than the DX station. For bands lower than 10 MHz, you transmit LSB on a frequency that is a few kHz lower than the DX station. The operation is the same for CW, but the split offset is less. Usually, 1-2 kHz rather than the 5-10 kHz offset used for SSB. If you are the rare DX or DXpedition station, you would, of course, reverse the split.

You can operate Split on digital modes, but it is less common. WSJT does it automatically for FT8 operation, to avoid problems caused by harmonics of the modulating audio signal.

Tuning in the split mode couldn't be easier! The main VFO knob tunes VFO-A (the receiver frequency) and the VFO ring tunes VFO-B (the transmitter frequency).

TIP: unlike the FTDX101, there is a TXW button that you can press to hear what is on the proposed transmitter frequency.

Pressing the SPLIT button moves the transmitter to the alternate VFO frequency, usually VFO-B. But it does not change the frequency or the mode. You need to be very careful because you could transmit on another band or the wrong mode.

Holding the SPLIT button until a double beep is heard can have two different effects depending on the QUICK SPLIT INPUT menu setting. Normally, with 'Quick Split' OFF, it sets VFO-B to the same mode and frequency as VFO-A, plus or minus the shift set by the QUICK SPLIT FREQ menu.

Press and holding the SPLIT button a second time moves the split offset further away from the receiver frequency by the same amount. For example, if the offset is +5 kHz, the transmitter frequency will move up 5 kHz each time you press and hold the SPLIT button. I am not sure that this is particularly useful since you can easily adjust the VFO-B frequency with the VFO tuning ring.

With QUICK SPLIT INPUT set to ON, holding down the SPLIT button brings up a popup window where you can set the split offset that you want to use. For example, if you want a five-kilohertz split, touch 5 then kHz. If you want a minus two kilohertz split touch -, 2, kHz.

<FUNC> <OPERATION SETTING> <GENERAL> <QUICK SPLIT INPUT>

<FUNC> <OPERATION SETTING> <GENERAL> <QUICK SPLIT FREQ>

For more information on how to use the Split mode effectively, see 'Operating Split' in the chapter on 'Operating the radio,' page 83.

MODE

Pressing the MODE button opens a popup menu so that you can change modes. You have to be quick because the popup only stays visible for about four seconds. The PRESET icon turns blue when the PRESET is active.

Press and holding the MODE button takes an 800x480 screenshot of the touchscreen display and saves it as a .bmp file to the 'FTDX10\Capture' directory on the SD card. For more information, see 'Changing the mode' on page 64.

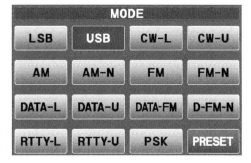

Figure 47: Mode selection

TIP: you can also activate the mode popup window by touching the mode icon to the right of the meter, on the touchscreen.

TXW

The TXW button temporarily changes the receiver to the same frequency as the transmitter. This is very handy if you want to quickly check that your split frequency is clear before you transmit. The button does not latch, you have to hold it down. The TXW button only works in Split mode. It has no effect if you use the receiver or transmit clarifier instead of the SPLIT button. Unlike some other radios, the TXW button does not open the receiver squelch.

ZIN/SPOT

ZIN is an auto-tune feature that pulls the receiver frequency onto a CW signal by matching the CW note with the Keyer Pitch. 'Netting' the receiver to the transmitter.

While receiving a CW signal, press ZIN/SPOT to pull the receiver VFO so that the received tone is the same as the keyer pitch. At that point, the transmitter will send on the same frequency as you are receiving.

Most of the time it works very well, although you may have to press the button two or three times to get exactly on frequency. The ZIN/SPOT button is only active in CW mode.

If you press and hold the ZIN/SPOT button the 'SPOT' keyer pitch tone will be heard. You can compare the tone of the incoming signal with the keyer pitch and tune the VFO and 'zero beat' the audio from the two sources.

LOCK

The LOCK button is located at the bottom right of the VFO cluster. It locks the VFO tuning and the VFO tuning ring (if it is set for tuning), so you don't accidentally bump the radio off frequency. This is quite easy to do when you are using the buttons clustered around the VFO knob.

The LOCK function is confirmed with a LED on the button and a 'LOCK' popup indicator appears you move the VFO or the VFO tuning ring. Press the LOCK button again to unlock the two VFO controls.

SHIFT

The SHIFT control moves the 24 kHz DSP filter passband by up to ±1.2 kHz to give you some ability to reject interference that is close to the wanted signal and inside the DSP passband. Shifting the I.F. is indicated on the filter function display. The shaded area indicates the R.FIL roofing filter passband and the white lines indicate the shifted I.F passband. A popup displays the amount of shift you have selected as you make adjustments. A LED next to the knob indicates that the I.F. shift is no longer zero.

Push and hold the SHIFT knob to reset both the I.F. Shift and the I.F. Width.

Figure 48: I.F. DSP controls

WIDTH

The WIDTH control changes the 24 kHz second I.F. bandwidth in the DSP stage of the receiver. A LED next to the knob indicates that the I.F. width has been adjusted.

On SSB, it is adjustable from 300 Hz to 4 kHz. The default is 3 kHz to match the bandwidth of the roofing filter.

On the CW, PSK, RTTY, and Data modes it is adjustable from 50 Hz to 3 kHz. The default is 500 Hz to match the bandwidth of the roofing filter.

The DSP bandwidth is fixed at 9 kHz on AM and 16 kHz on FM.

The current DSP channel bandwidth is shown graphically on the filter function display below the VFO frequency display. The shaded area indicates the R.FIL roofing filter passband and the white lines indicate the adjusted I.F. passband. As you adjust the WIDTH control a popup shows the I.F. bandwidth in Hertz.

Push and hold the SHIFT knob to reset both the I.F. Shift and the I.F. Width.

If you are still experiencing interference within the DSP bandwidth, even after employing the I.F. shift and width controls, you can employ the digital notch filter, the manual notch filter, or the contour control.

NOTCH

Turning the NOTCH knob automatically turns on the NOTCH filter button. An orange 'V' indicating the notch appears on the filter function display. Turning the knob moves the notch frequency across the audio spectrum and displays the audio frequency on a popup display. I normally tune the control to the right spot by listening to the interfering signal disappear. You can achieve the same thing by moving the notch onto the interfering signal indicated on the filter function display.

Once you have adjusted the notch frequency you can toggle the filter on and off using the NOTCH button. Press and holding the NOTCH button turns resets the notch frequency to the center of the filter passband.

You can change the width of the manual notch filter. The NARROW setting will be fine for single carrier interference, and it will have less impact on the quality of the wanted signal. The default WIDE setting is better at nulling out wider signals that are modulated with noise or narrowband data.

<FUNC> <OPERATION SETTING> <RX DSP> <IF NOTCH WIDTH>

TIP: if the interfering signal is very strong, Yaesu recommends using the manual filter before trying the DNF (digital notch filter). This is because the manual notch is deeper than the auto-notch.

CONT/APF CONTROL

The contour control was new to me, and I quite like it. It produces a small dip or boost in the audio frequency response which you can move across the audio passband from 50 to 3200 Hz. It's like a 'mini-notch filter.' You can hear the effect as you tune across the audio signal. It can attenuate an interfering signal without notching out the audio frequency completely or it can be used as a sort of tone control, attenuating, or boosting the low or high audio frequencies.

Turning the CONT knob (outside ring of the NOTCH knob) automatically turns on the CONT button. An orange 'dip' or 'bump' indicator is shown on the filter function display. Once you have adjusted the contour frequency you can toggle the filter on and off using the CONT button. Press and hold the CONT button to reset the contour filter to the center of the passband.

TIP: contour adjustment of the higher audio frequencies has little effect on CW because the 500 Hz roofing filter attenuates those frequencies anyway. If you have a hearing impairment, for example, hearing loss at high frequencies you could set the contour level to a positive value and use the contour function to boost the high audio frequencies.

You can change the bandwidth and the gain of the contour filter. You can even make it a hump so that it acts even more like a tone control, amplifying a part of the audio bandwidth. As an experiment, I set the control to +18 and could clearly see the increase in noise level as I tuned the contour control across the audio band.

<FUNC> <OPERATION SETTING> <RX DSP> <CONTOUR LEVEL> (-40 to +20)

<FUNC> <OPERATION SETTING> <RX DSP> <CONTOUR WIDTH> (1 to 11)

The default level is -15 dB and the default width is 10.

The APF (audio peak filter) button is for CW only. It is disabled on all other modes. Tune in a CW signal by ear or by using ZIN or the audio bar-graph indicator and turn on APF. The very narrow audio filter eliminates nearly all of the band noise revealing a clean CW signal. "It is fab!" The APF filter frequency is adjustable.

Turning the CONT/APF knob while in the CW mode will activate the CONT/APF button and enable the knob. APF is adjustable over a range from -250 Hz to + 250 Hz. Press and hold CONT/APF to reset the offset to zero. You can also change the bandwidth of the APF filter. The narrow filter will provide the cleanest sounding CW signal, but the signal has to be exactly on frequency. The wide setting is easier to tune, but you will hear more band noise.

In the CW mode,

- Press the CONT/APF button for the APF filter. This is indicated with a vertical orange line next to the P marker on the filter function display. Use the CONT/APF knob to adjust the filter frequency.

- Press the CONT/APF button again for the Contour filter. Indicated with an orange dip or peak on the filter function display. Initially, it will be outside of the 500 Hz CW passband but you can use the CONT/APF knob to bring the filter frequency inside the passband.

- Press the CONT/APF button again to turn off both filters.

In all other modes,

- Press the CONT/APF button for the Contour filter. This is indicated with an orange dip or peak on the filter function display. Use the CONT/APF knob to adjust the filter frequency.

- Press the CONT/APF button again to turn off both filters.

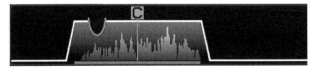

Figure 49: Contour filter

Figure 50: APF filter

Front panel connectors

PHONES

The PHONES jack is for connecting your stereo headphones. It takes a standard 3.5 mm (1/8") **stereo** Phone plug. The receiver audio is on both stereo channels. To avoid a loud "pop" in your ears, put your headphones on, after turning on the radio.

MIC JACK

The 8 pin Mic jack is for the microphone. It uses an RJ45 connector similar to many mobile radios, rather than the traditional 8 pin connector. The rubber boot offers some protection against dust.

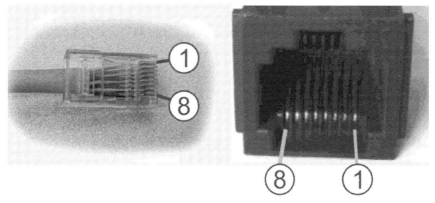

Figure 51: Microphone jack connections

1	Down button (DWN)	5	Mic audio
2	Up button (UP)	6	PTT
3	+5V for Electret microphones	7	Ground
4	Mic Ground	8	Fast button

The SSM-75E microphone supplied with the radio has seven buttons in addition to the PTT switch. The UP and DWN buttons on the top of the microphone step the VFO frequency using the same step size as the VFO. The FAST button multiplies the step size by ten. In memory channel mode the UP/DWN buttons step through the memory channels. Press and hold to start a memory channel scan.

The MUTE button mutes the receiver audio. Very handy if someone comes into the shack while you are operating. But the button does not latch. You have to hold it down.

P1 locks and unlocks the VFO, the same as pressing the LOCK button.

P2 cycles through the QMB (quick memory bank) channels. Press and hold saves the current frequency and mode to the QMB memory. The same as using the QMB button.

P3 swaps the VFOs, the same as pressing the A/B button. (Handy for split operation). Press and hold makes the sub VFO the same as the current VFO.

P4 swaps between VFO and memory channel mode, the same as pressing the V/M button.

SD CARD SLOT

The SD card is used to hold a backup of the radio's current menu settings and memory channels, and for saving screenshots, message keyer macros, and off-air recordings. It is also used for firmware updates.

You can use a 'full size' SD card, or a micro SD card in an adapter. Many micro SD cards are sold with a free SD adapter. The card can be a 2 Gb SD card or any SDHC card from 4 Gb up to 32 Gb. I am using a 16 Gb SanDisk SDHC card. The information screen says that I have used 100 Mb so far, so buying a 16 Gb card like I did, is probably overkill. I leave the SD card in the radio all the time except while transferring files to or from my computer.

<FUNC> <EXTENSION SETTING> <SD CARD> <Informations DONE>

TIP: don't pay extra for a fast 'Extreme' or 'React+' SDHC card. This is not an application where speed is important. Buy a slower / cheaper 'Ultra' or 'Select+' card. If you buy a micro SD card, make sure you get one that comes with the plastic micro SD to full-size SD adapter.

Yaesu recommends formatting the SD card before using it for storage <FUNC> <EXTENSION SETTING> <SD CARD> <Format DONE>. I didn't bother because the card has FTDX101 files on it, and I have not experienced any problems.

The micro SD card slides into the adapter, contacts first, with the contacts facing down, the same side as the contacts on the adapter. The card will only fit one way. Slide the full-size SD card, or micro SD card and adapter, into the SD card slot on the radio. Contacts first, with the contacts facing down. The cut-off corner should be on the right. The card will only fit one way. It should slide in easily and latch with a click.

If the radio is turned on when you insert the SD card you will be presented with a popup menu that offers a 'Setup?' option (YES/NO).

Selecting YES is perfectly safe. It just opens the <FUNC> <EXTENSION SETTING> <SD CARD> setup screen. Selecting NO closes the popup window.

To remove the SD card. Make sure that the transceiver is not writing data to the card, then gently push the card to un-latch it and remove your finger.

The card should pop out far enough for you to be able to pull it out of the radio. It should slide easily. If it resists, repeat the 'push to unlatch' process.

The radio will automatically add subdirectories to the SD card.

- Capture holds .bmp screen capture images.

- MemList holds memory channel backup files.

- Menu holds menu system backup files.

- MESSAGE holds the voice keyer recordings as .wav files.

- PlayList holds off-air recordings as .wav files.

- The root directory holds firmware update files. They **must** be in the root directory, not a subdirectory. The overall "firmware version" is the file marked FTDX_MAIN. In this case V01-08.

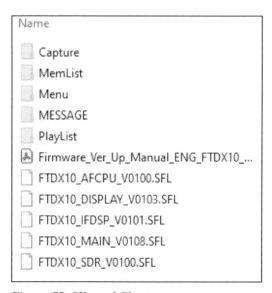

Figure 52: SD card file structure

Rear panel connectors

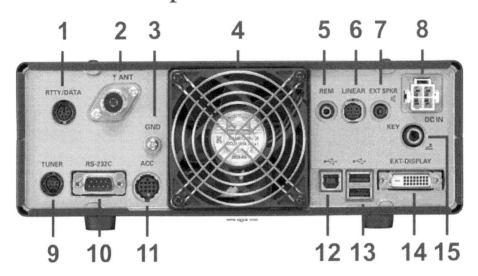

Figure 53: FTDX10 rear panel

Image: RigPix Database - Yaesu - FTdx-10

1	RTTY/Data (6 pin mini DIN)	9	Tuner (8 pin mini DIN)
2	ANT (antenna jack)	10	RS-232 serial (DB-9)
3	Ground lug	11	ACC (13 pin DIN)
4	Fan	12	USB (Type B)
5	REM (remote)	13	USB 2.0 type A (2 jacks)
6	Linear amp jack (10 pin mini DIN)	14	External Display (DVI-D)
7	External speaker jack (mono)	15	Key
8	DC input		

1. RTTY/DATA

The RTTY/Data jack is for connection to a TNC for digital mode operation. These days you are more likely to use the USB cable unless you already have a TNC setup from a bygone era. It can be used for CW keying or HF Packet, but it is most commonly used for RTTY. There are pins for audio data input and output, PTT, ground, Squelch, and 'SHIFT' which is the FSK (keying data) input.

TIP: The plug is a 6 pin mini-DIN connector. I have found the easiest way to source one is to chop the mouse off an old PS2 mouse and use the cable. If you don't have a PS2 mouse, try your local op-shops they often have a few because nobody buys them anymore.

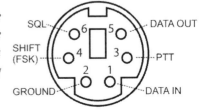

2. ANT

ANT is the antenna jack used for receiving and transmitting. The connector is a standard 50 Ω, SO-239 UHF jack. It takes a PL-259 plug.

3. GND

The radio ground connection should be connected to your shack ground in a 'star' rather than daisy chain format. The shack ground should be connected to an earth stake or earth mat outside. NOT to the mains earth. Earthing the radio can protect the radio from lightning static discharge (not direct lightning strikes). It can also improve noise performance.

4. COOLING FAN

The fan is important to stop your radio from overheating on long transmissions. Make sure that it is not obstructed.

5. REM

The REM mono mini phone jack is used to connect the FH-2 keypad. You can buy one at a reasonable price or construct your own with a few push buttons and resistors. I found this schematic on the Internet.

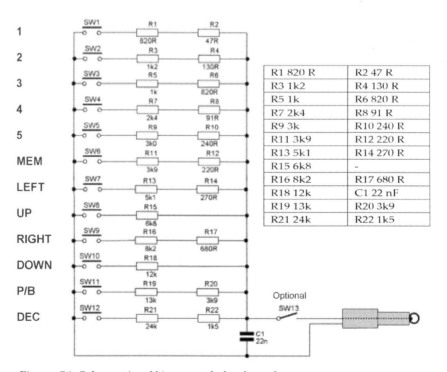

R1 820 R	R2 47 R
R3 1k2	R4 130 R
R5 1k	R6 820 R
R7 2k4	R8 91 R
R9 3k	R10 240 R
R11 3k9	R12 220 R
R13 5k1	R14 270 R
R15 6k8	–
R16 8k2	R17 680 R
R18 12k	C1 22 nF
R19 13k	R20 3k9
R21 24k	R22 1k5

Figure 54: Schematic of Yaesu style keyboard

6. LINEAR

The 10 pin mini-DIN connector is used to provide band switching information, PTT, amplifier inhibit, and ALC information to the Yaesu VL-1000 linear amplifier. The band data can be used by other compatible amplifiers, or possibly for an external antenna tuner, or for preamplifier band switching.

The FTDX10 does not have RCA jacks for linear amplifier PTT and ALC. Use pin 9 for the amplifier PTT, pin 10 for common ground, and pin 2 for ALC input to the radio.

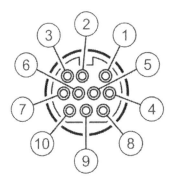

1. Transmit inhibit
2. ALC input
3. TX Request input
4. Band data A
5. Band data B
6. Band data C
7. Band data D
8. 13.8V out (200mA)
9. TX Ground
10. Ground

Figure 55: Linear amp jack pinout

➢ **Yaesu band switching**

Band switching data is available on the 10 pin connector. It uses the standard BCD (binary coded decimal) Yaesu format.

Band	Frequency	DCBA	Hex
160m	1.8 MHz	0001	01
80m	3.5 MHz	0010	02
60m	5 MHz	0011	03
40m	7 MHz	0011	03
30m	10 MHz	0100	04
20m	14 MHz	0101	05
17m	18 MHz	0110	06
15m	21 MHz	0111	07
12m	24.8 MHz	1000	08
10m	28 MHz	1001	09
6m	50 MHz	1010	0A

7. EXT SPEAKER

The 'high level' mono speaker jack on the rear of the radio is for connection to an external speaker. If you want to use a set of amplified stereo PC speakers, use the front panel headphone jack instead. The speaker jack is a mono 3.5 mm (1/8") phone connector.

TIP: the headphone jack is a stereo jack with receiver audio on both channels. If you plug a stereo plug into the Ext Speaker jack you will only get sound on the left side.

8. DC INPUT

The FTDX10 requires a 13.8 volt (± 15%) regulated power supply capable of supplying at least 23 Amps (25A recommended). Use the supplied power lead. Never use a power lead without inline fuses.

BE VERY CAREFUL TO SUPPLY THE CORRECT DC POLARITY. Red is positive and black is negative.

The connector has a locking tab on the top. It may be covered by black plastic connector protection. Squeeze the tab down before attempting to unplug the DC cable.

9. TUNER

The 8 pin mini-DIN 'Tuner' jack is for connecting a Yaesu FC-40 or Yaesu compatible antenna tuner. Use the cable and ferrite isolator supplied with the tuner. The pinout is in the Yaesu manual (page 13).

10. RS-232

The RS-232 port can be used for CAT control and RTS/DTR signaling if you want to go "old-school," and if you can find a computer that has an RS-232 interface. The connector is a standard DB-9 serial interface. "Get into the nineties, man!" Use a USB cable.

11. ACC

The ACC (Accessory) jack is a 13 pin standard size DIN connector. It could be used for transceiver control from a variety of external hardware devices. It is configured for connection to the Yaesu SCU-LAN10 Ethernet remote control interface. The pinout is on page 14 of the Yaesu manual.

12. USB TYPE B PORT

The USB Type B connector is used for CAT control of the transceiver and for transferring audio to and from a PC for external digital mode software. See 'Setting up a USB connection' on page 32.

You will need a 'Type A to Type B, USB 2.0' cable to connect the radio to a PC.

13. USB TYPE-A PORTS

The USB Type-A connectors are used to connect a mouse or a keyboard to the radio.

The reason that the rear panel of the radio has a Type-B USB port and two Type-A USB ports, is that they have different functions. USB Type-A ports are "host" ports used for connecting USB peripheral devices such as a keyboard or a mouse. USB Type B ports are used when a peripheral, in this case, the transceiver, is connected to a remote host such as your PC. USB Type-B connectors are used on items like scanners or printers.

➢ **USB mouse**

Connecting a computer mouse is easy. Just plug a USB mouse into one of the front panel USB ports and you are ready to go. I tried three wireless mice and none of them worked, but apparently, the Logitech M310 and M305 wireless models will work. The right mouse button has no function and sadly the mouse wheel does nothing either. It could have been used to scroll the frequency… but no.

So, what can you do with the mouse?

- The left mouse click has the same effect as touching the screen with your finger. You can change the meters, the filter function display, the waterfall to spectrum ratio, the VFO frequencies, and operate the Soft Keys.

- If you press the FUNC knob to open the FUNC menu, you can use the mouse to select any of the menu functions. But there is no way to use the mouse to open the FUNC menu.

If you use the external monitor, you will probably use a mouse to control the screen.

➢ **USB keyboard**

You can connect a USB keyboard to the radio and use it to fill in the TEXT messages. However, I can't see much point in using an external keyboard when there is an onscreen keyboard provided for this purpose. Unfortunately, you cannot use an external keyboard in conjunction with the digital mode decoders to send text directly from the keyboard.

TIP: Another odd feature is that if I plug in my Logitech wireless keyboard and mouse combo. The keyboard works, but the mouse doesn't.

14. EXT DISPLAY

The External Display port is a DVI-D connector for the connection of a computer monitor. You can buy a DVI-D to HDMI adapter if your display has an HDMI input but no DVI-D input. The monitor must support 800x480 or 800x600 resolution. That could be a problem. My displays do not support 800x480. They will only go down to 800x600. If I select the 800x480 pixel option on the radio I just get a larger picture.

You must change a menu setting to use an external monitor.
<FUNC> <DISPLAY SETTING> <EXT MONITOR> <EXT DISPLAY> ON

And I recommend selecting the higher screen resolution option.
<FUNC> <DISPLAY SETTING> <EXT MONITOR> <PIXEL> 800x600

While 800x600 pixels looks great on the five-inch touchscreen, it looks pretty bad on a modern 1900x1200 monitor. It reminds me of the Lego movie. You get a big, low-resolution, copy of the touchscreen. The monitor display has no touchscreen capability, but you can control the screen with a USB computer mouse.

The external display is not equivalent to the display offered by an SDR program. Although you can use a mouse to operate any of the touchscreen controls, you cannot control the functions of the physical buttons and knobs.

The external display might be useful if you are demonstrating the radio at a public event, or you have poor eyesight. But I don't see any appeal in trading the excellent touchscreen display for a big blocky image on a monitor.

15. KEY

The CW key jack is a ¼″ (6.35 mm) stereo Phone jack. It can be used for a straight key or a paddle. The 'key up' voltage is 5.0 volts. The 'key down' current is 3 mA.

Paddle — DOT DASH COMMON   Straight key — KEY

Troubleshooting

Most of the items covered in the troubleshooting chapter are not faults. However, they are conditions that might worry you if you encounter them while operating the radio. This chapter may help you if something unexpected happens. The items are not listed in any particular order, just the way I discovered them.

YAESU SERIAL NUMBER IDENTIFICATION

It may be of interest to identify the date range of your Yaesu transceiver.

Digit	Function	Range
1	Last digit of the year of manufacture	0 = 2020 1 = 2021 2 = 2022 3 = 2023 etc.
2	Month of manufacture	C=Jan, D=Feb, E=Mar, F=Apr, G=May, H=Jun, I=Jul, J=Aug, K=Sep, L=Oct, M=Nov, N=Dec
3,4	Lot number	01 to 99
5,6,7,8	Serial number	0001 to 9999

For example,

1E041299	=	2021	March	Lot 04	Serial 1299
1L010001	=	2021	October	Lot 01	Serial 0001

BLACK SCREEN AND A YAESU LOGO ON THE SCREEN

Don't panic. The radio has a screen saver function to preserve your display just like your PC does. Touch the screen or press any button to restore normal operation. You can turn it off, but I don't recommend doing that. The time before the screen saver starts is adjusted using <FUNC> <DISPLAY SETTING> <DISPLAY> <SCREEN SAVER>. You can choose from OFF, 15 minutes, 30 minutes, or 60 minutes.

WATERFALL TOO DARK.

Only large spectrum peaks are showing on the spectrum display. This occurs when the spectrum level is too low. Perhaps you changed to a narrow band mode, a quieter band, or a different antenna. Select <FUNC> <LEVEL> and turn the FUNC knob clockwise to increase the spectrum level until the spectrum noise floor is just visible at the bottom of the window. The waterfall brightness should now be correct.

You can set the VFO tuning ring (MPVD) to LEVEL by touch and holding the CS button and leaving the CS button turned on.

WATERFALL TOO LIGHT.

This happens when the spectrum level is too high. Usually, if you change to a wide band mode, a noisier band, or select a different antenna. Select <FUNC> <LEVEL> and turn the FUNC knob anti-clockwise to decrease the spectrum level until the spectrum noise floor is just visible at the bottom of the window. The waterfall brightness should now be correct. You can also set the VFO tuning ring (MPVD) to LEVEL by touch and holding the CS button and leaving the CS button with its LED turned on.

HARDWARE RESET.

Before resorting to a hardware reset, try a software reset. See 'Extension Settings - Reset' on page 150. If that fails, try a 'cold boot.' Turn off the radio and turn off, or disconnect, the DC power supply. Then reapply the DC and turn on the transceiver.

If the software reset and the 'cold boot' don't work and the radio is still dead, you can try a hardware microprocessor reset. Turn the transceiver off, then turn it on again while holding down the FINE and LOCK buttons. When the transceiver starts, release the two buttons.

DATE OR TIME SLIP

Date or time slippage may possibly be caused by some special relativistic 'time dilation' due to traveling at extreme speeds, or it might be due to the slightly odd way the date and time are set. The latter issue happened to me. I noticed that the radio was date stamping saved screenshot files with yesterday's date.

Press <FUNC> <EXTENSION SETTING> <DATE&TIME> and set the Day, Month, and Year. **Then press <FUNC> to save the date**.

Then you can set the HOUR, & MINUTE and **press <FUNC> to save the time**.

If you do not press FUNC both times, the date and/or time will slip.

Tip: If you want your files date stamped with UTC date and time, enter UTC date and time details. But I believe it is easier to use local time.

VOICE MEMORY KEYER WON'T TRANSMIT

You must have BK-IN turned on to transmit the voice keyer messages. However, you can listen to your recordings by keying the messages with BK-IN turned off.

As of firmware revision V01-08, the voice keyer will not transmit if PRESET is on. Even if you have changed from the DATA-U mode to SSB. The transmitter will turn on, but no power will be generated on SSB. Make sure that PRESET is off. Press or select MODE and make sure PRESET is grey.

A/B BUTTON BEHAVING BADLY

You expect the A/B button to toggle the radio between VFO-A and VFO-B. However, if either VFO is in the memory channel mode, pressing A/B toggles between the MEM mode and the VFO. If both VFOs are in the memory channel mode, pressing A/B appears to do nothing, but it is actually swapping between the two memory channels. I guess that could be handy if you want to toggle the radio between two memory channels.

Press A/B to show the two LEDs on MEM. Press V/M to return the VFO to normal. Press A/B to swap VFOs. If the LEDs go back to MEM, press V/M to return the VFO to normal. Now the A/B button should toggle between VFO-A and VFO-B.

SPLIT WEIRDNESS

Split operation can be a bit odd if you have been using the memory channels. If VFO-B has been left in the MEM (memory channel mode) and you press SPLIT the radio will transmit on the MEM frequency, not the VFO-B frequency. You can see this on the LED panel.

It is the same but reversed if VFO-A has been left in the memory channel mode. When you press SPLIT the radio will receive on the MEM frequency and transmit on the VFO-B frequency. See, 'A/B button behaving badly' above.

ENABLE THE 60M AND 4M BAND FREQUENCY ALLOCATION

I have not tried this, so it is at your own risk if you do. However, the information came from a "Yaesu UK" YouTube video, so it should be OK.

This process resets your radio settings so you should back up the menu and memory lists first and restore them after you have changed the region.

If you are operating the radio in the UK and wish to access the UK 5 MHz (60m) and 70 MHz (4m) frequency ranges, 5.250 MHz - 5.4065 MHz and 70.000 MHz - 70.499 MHz.

1. Turn off the transceiver.

2. Hold down the FINE and MODE buttons while you restart the radio.

3. The VFO-A RX LED should flash.

4. Press FUNC. The VFO-A RX LED should flash more slowly.

5. Press and hold FUNC. The radio should re-boot with the UK band settings.

It is your responsibility to adhere to your amateur radio license conditions and only receive and transmit in frequency ranges in accordance with your amateur radio license. Search for How to change the region on the FTDX10 to UK - YouTube.

FILTER FUNCTION DISPLAY NOT SHOWING SPECTRUM

Touch and holding the filter function display turns off the spectrum display, leaving only the passband indicator and the notch or contour indicators if they are turned on. Touch and holding the filter function display again should restore the spectrum display.

SHIFT OR WIDTH CONTROL WON'T RESET

Pressing the SHIFT knob results in three beeps, indicating that the function is not available. To reset the I.F. Shift and Width, you have to press and hold the knob until you hear two beeps. Why not just clear them on a simple press?

DIGITAL MODE AUDIO

The audio signal is sent over the USB cable to the PC in any mode. So, you can use your digital mode PC software to see and decode digital mode signals like RTTY, PSK, or FT8. But you cannot transmit digital mode audio signals unless you are in a data mode, usually DATA-U. *Use DATA-FM for FM Packet radio on 6m.* CW is digitally keyed, so it will work when the transceiver is in CW mode. FSK RTTY is also digitally keyed, so you should use the RTTY-L mode for that. Note that not all PC digital mode software supports FSK keying.

DIGITAL MODE IS NOT THE SAME

Your external digital mode was set up and working fine, but now the transmit power is low, or the received signals look different on the PC software, or it just won't transmit at all. Check that you have turned on the Preset. Or turned off if you were not using it last time. The Preset function changes many of the default settings. Touch and hold the PRESET Soft Key, to enter the Preset setup menu.

NO AUDIO TO EXTERNAL DIGITAL MODES PROGRAM

The transceiver volume control does not affect the audio level sent to the computer over the USB cable, but the squelch control does. It must be open for audio to be sent to an external program. This is rather annoying. I wish that it didn't affect the audio to the PC, but I guess you can always turn down the volume.

S METER STUCK AT A HIGH LEVEL

The S meter being fixed at a high level and not moving is an indication that the RF level has been turned down. Turn the RF gain up to the maximum or change the RF/SQL VR control to Squelch instead of RF gain.

<FUNC> <OPERATION SETTING> <GENERAL> <RF/SQL VR> (RF or SQL)

The other possibility is that the radio is transmitting. Check that the red TX icon on the LED display is not active.

TRANSCEIVER STUCK IN TRANSMIT MODE

➢ **Transmitter held on by the USB or RTTY/DATA COM port**

Symptom: starting your PC digital mode software causes the transceiver to start transmitting.

This is not a fault, but it can be rather disconcerting. In SSB or DATA-U mode it is unlikely to damage anything as there will be very little power transmitted. However, in CW AM or FM mode, it will cause full power to be transmitted.

The problem is usually due to the RTS / DTS settings on the COM port being inappropriately controlled by the PC software. Normally the transceiver is set so that the RTS signal is PTT (turns transmit on) and the DTS signal is used to send CW or RTTY data. If the digital mode software has either RTS or DTR held 'active' or 'low' it can make the transceiver switch to transmit. If this happens, check the COM port settings in the digital mode software and make sure that the RTS and DTR lines are set the same way as the transceiver. This condition can be proved out of the radio by removing the USB cable or the RTTY/DATA cable if that's what you are using.

TIP: the RTS and DTR lines on the Enhanced COM port are not used for keying and can be set to unused, high, or off. The RTS/DTR lines on the Standard COM port are used for keying. They should be set the same as the software. Typically, RTS for PTT and DTR for CW or FSK keying.

It is usually possible to set the digital mode software to use a CAT command for PTT rather than the RTS (or DTR) line. But use RTS/DTR signaling if you can.

➢ **Other reasons the transmitter may be stuck on.**

Check that the front panel MOX button has not been turned on. There is no LED indication other than the TX LED. It could also be the Morse key or the microphone PTT. Check that the Microphone PTT button is not stuck on transmit, unplug the Morse key, and remove the control cables from the rear panel LINEAR, RTTY/DATA, and USB jacks. A stuck key on the external FH-2 keypad is also a possibility.

Basically, unplug all the external cables except the antenna and the DC supply until the problem is resolved. Then plug them in one by one to problem-solve which cable connection is causing the issue.

If you still have a problem. Turn off the power using the power switch. When the radio re-starts it will hopefully be back to normal.

MEMORY CHANNELS WILL NOT SELECT

There is some general weirdness about the memory channel selection if 'Groups' is turned on. <FUNC> <OPERATION SETTING> <GENERAL> <MEM GROUP>.

Constraint number 1: You cannot use the <FUNC> <M-GROUP> menu item to change between groups unless the radio is in memory channel mode (V/M). In VFO mode you can only display the current group setting. This is mildly annoying and to my mind an unnecessary constraint.

Constraint number 2: If you have Groups turned on, the STEP/MCH and microphone UP and DWN buttons can only select or scan memory channels within the currently selected group. In other words, you have to; select Memory Channel mode with V/M, then select the group with <FUNC> <M-GROUP>, then press and hold STEP/MCH, before you can use the VFO tuning ring to step through the memory channels.

While having groups is a great idea, the difficulty of choosing memory channels makes using them a challenge. I am not sure that it is worth the hassle.

CW MODE - SIDETONE BUT NO TRANSMITTER POWER

This one has caught me out several times. CW will not be transmitted unless break-in has been turned on by pressing <FUNC> <BK-IN>.

Alternatively, with BK-IN turned off, pressing MOX, pressing the PTT button on the microphone, sending a CAT command from PC software, or switching the PTT line using the PTT line on the RTTY/DATA connector will put the transmitter into transmit mode and the CW signal will be transmitted.

CW MODE – NO SIDETONE

You have to have the MONI LEVEL turned up to hear the sidetone. MONI is indicated above the spectrum display.

ANTENNA TUNER DOES NOT WORK ON 70 MHZ BAND

As far as I know, the tuner does not work on the 70 MHz, 4m band. My radio is from the US market and the 70 MHz (4m band) is not a Ham band in the USA or New Zealand for that matter. It may work on UK region radios. On my radio, the tuner only works from 1.8 MHz to 29.7 MHz and on the 6m band.

USB MOUSE WILL NOT WORK

Many USB mice will not work. Apparently, the Logitech M310 and M305 wireless models will work. Or you can use a 'wired' USB mouse with a cable. You will probably want to use a mouse if you are using an external monitor. I find the external display so awful, and mouse operation so limited, I don't bother using one.

CALIBRATING THE SPECTRUM SCOPE FOR TRANSMITTING

I expect that few people will bother calibrating the spectrum scope for transmitting because there isn't a separate LEVEL adjustment for transmitting ☹. As soon as you go back to receiving you will probably adjust the spectrum scope level and the calibration will be lost.

Select the CW mode with the keyer turned off and BK-IN turned on. Key the transmitter with your Morse key and use the LEVEL control to set the top of the CW signal to the top of the spectrum display. That calibrates the top line of the spectrum scope to +50 dBm (100W) PEP or +53 dBm (200W) on the MP. The spectrum scope is factory calibrated to 5 dB per division so when the top line is calibrated to +50 dBm, each line up from the bottom represents a 5 dBm increase in transmitter power. Now return to SSB. When you talk, the IMD (intermodulation distortion) products should not exceed +20 dBm [+23 dBm on the MP]. (30 dB below the PEP level). Put simply, the IMD products should not exceed the bottom four lines of the calibrated display. +23 dBm is 0.2 Watts. +20 dBm is 0.1 Watts.

CAT CONTROL NOT WORKING

Some PC software does not like having CAT RTS (interrupt control) turned on. Neither WSJT-X nor MixW will control the radio with CAT RTS turned on. The preset turns it off by default but the FUNC menu has it turned on by default.

<FUNC> <OPERATION SETTING> <GENERAL> <CAT RTS> <OFF>

Note: This relates to the RTS setting for the 'Enhanced' Com port, not the RTS used for PTT signaling on the 'Standard Com port.'

REFERENCE OSCILLATOR FINE ADJUSTMENT

I don't think that it is necessary to adjust the reference oscillator and if you do, it will probably drift back again as soon as the transceiver changes temperature when you transmit, or over time. The FTDX10 has a reference oscillator with a stability of ± 0.5 ppm. This is much better than most amateur radio transceivers. It means that it should be within ± 15 Hz when the radio is tuned to 30 MHz. To adjust the reference oscillator: Let the transceiver run on receive for at least 30 minutes so that the temperature inside the radio has stabilized. Select CW mode and tune to 15.000000 MHz or 10.000000 MHz to receive the WWV/WWVH carrier signal. You should hear an audio tone equivalent to the <FUNC> <CW Pitch>. Press and hold down the ZIN/SPOT button and you should hear a slow beat note between the received tone and the CW pitch tone. Select <FUNC> <OPERATION SETTING> <GENERAL> <REF FREQ FINE ADJ> and move it plus 1 Hz or minus 1 Hz. Repeat the test holding down the ZIN/SPOT button. If the beat note is faster, adjust the reference frequency back in the other direction. You are aiming for a very slow beat note. See the N4HNH video at https://www.youtube.com/watch?v=0ClYJ6tgRnY.

Glossary

Term	Description
3DSS	The 'three dimensions signal stream' is a spectrum display that includes signal history similar to the information on a waterfall display. It is represented as a 3D image tapering back to give an illusion of depth.
6m, 10m 20m, 40m, 80m	50 MHz, 28 MHz, 14 MHz, 7 MHz, and 3.5 MHz amateur radio bands
59	Standard (default) signal report for amateur radio voice conversations. A report of '59' means excellent readability and strength.
599, 5NN	Standard (default) signal report for amateur radio CW conversations. A 599 report means perfect readability, strength, and tone. The 599 signal report is often used for digital modes as well. The 5NN version is faster to send using CW. It is often used as a signal report when working contest stations.
73	Morse code abbreviation 'best wishes, see you later.' It is used when you have finished transmitting at the end of the conversation.
.dll	Dynamic Link Library. A reusable software block that can be called from other programs.
A/D	Analog to digital
ACC	13 pin DIN accessory jack
ADC	Analog to digital converter or analog to digital conversion
AF	Audio frequency - nominally 20 to 20,000 Hz.
AFSK	Audio Frequency Shift Keying. An RTTY mode that uses tones rather than a digital signal to drive the SSB transmitter.
AGC	Automatic Gain Control. In the FTDX10 AGC is performed in the I.F. DSP stage.
ALC	Automatic Level Control. There are two kinds used by the radio. One is the AMC (automatic microphone gain) used to ensure that the radio is not overmodulated. This is metered by the COMP meter. Confusingly the ALC meter measures the MIC GAIN. The other form of ALC is the negative control voltage sent from a linear amplifier to the ALC input on the LINEAR jack to ensure that the amplifier is never overdriven by the transceiver.
Algorithm	A process, or set of rules, to be followed in calculations or other problem-solving operations, especially by a computer. In DSP it is a mathematical formula, code block, or process that acts on the

	data signal stream to perform a particular function, for example, a noise filter.
AM	Amplitude modulation, (double sideband with carrier)
ANT	Abbreviation for antenna
APF	Audio Peak Filter. This is a very sharp filter that operates on the DSP audio signal. It is only available in the CW mode. The APF frequency can be offset using the CONT/APF knob.
ATT	Abbreviation for attenuator. The FTDX10 has front-end attenuator selections for 6 dB, 12 dB, or 18 dB.
Band scope	A band scope is a spectrum display of the frequencies above and below the frequency that the radio is tuned to. The center of the display is generally the frequency that you are listening to. This is different from a 'Panadapter' spectrum and waterfall display where you can listen to any frequency across the display.
Bit	Binary value 0 or 1.
BK-IN	CW 'Break-in' is the practice of receiving Morse code between the Morse characters or words that you are sending. The radio will not automatically transmit in the CW mode unless BK-IN is turned on.
BPSK	Binary phase-shift keying. Digital transmission mode using a 180-degree phase change to indicate the transition from a binary one to a binary zero. The PSK decoder in the FTDX10 supports BPSK31 and QPSK.
BW	Bandwidth. The range between two frequencies. For example, an audio passband from 200 Hz to 2800 Hz has a 2.6 kHz bandwidth.
Carrier	Usually refers to the transmission of an unmodulated RF signal. It is called a carrier because the modulation process modifies the un-modulated RF signal to carry the modulation information. A carrier signal can be amplitude, frequency, and/or phase modulated. Then it is referred to as a 'modulated carrier.' An oscillator signal is not a carrier unless it is transmitted.
CAT	Computer-aided transceiver. Text strings are used to control a ham radio transceiver from a computer program.
CENTER	Sets the spectrum and waterfall display so that VFO frequency is in the center of the display with a span of frequencies on either side. It is a band scope display rather than a panadapter display.
CODEC	Coder/decoder - a device or software used for encoding or decoding a digital data stream.
COM	Serial communications port. In the FTDX10 the two Com Ports are 'virtual' serial ports for the data carried over the USB cable. There is also a standard RS-232 com port on the rear panel.

COMP	The meter position for measuring the speech processor (compressor) level when PROC LEVEL is turned up. It measures the AMC LEVEL when the speech processor is turned off.
CPU	Central processing unit - usually a microprocessor. Can be implemented within an FPGA.
CQ	"Seek You" is an abbreviation used by amateur radio operators when making a general call which anyone can answer.
CW	Continuous Wave. The mode used to send Morse Code.
D/A	Digital to analog.
DAC	Digital to analog converter or digital to analog conversion
data	A stream of binary digital bits carrying information
DATA	One of the data modes, DATA-U, or DATA-FM, used to interface the radio with a PC digital mode program. You must be in a DATA mode to transmit from a PC digital mode program.
dB, dBm, dBc, dBV	The Decibel (dB) is a way of representing numbers using a logarithmic scale. Decibels are used to describe a ratio, i.e. the difference between two levels or numbers. They are often referenced to a fixed value such as a Volt (dBV), a milliwatt (dBm), or the carrier level (dBc). Decibels are also used to represent logarithmic units of gain or loss. An amplifier might have 3 dB of gain. An attenuator might have a loss of 10 dB.
DC	Direct Current. You need a 25 Amp regulated 13.8V DC supply to power the radio.
Digital modes	Amateur radio transmission of digital information rather than voice. It can be text or data such as video, still pictures, or computer files, (PSK, RTTY, FT8, Olivia, SSTV, etc.)
DNF	The Digital Notch Filter automatically eliminates the effect of long-term interference signals such as carrier signals that are close to the wanted receiving frequency. Not effective against impulse noise.
DNR	Digital noise reduction. A DSP filter.
DSP	Digital signal processing. The FTDX10 uses a Texas TMS320C6746 DSP chip. DSP uses mathematical algorithms running in computer firmware to manipulate digital signals in ways that are equivalent to functions performed on analog signals by hardware mixers, oscillators, filters, amplifiers, attenuators, modulators, and demodulators.
DTR	'Device Terminal Ready' is a com port control line often used for sending CW or FSK RTTY data over the CAT interface between the radio and a PC.
DVI-D	Digital Visual Interface is the video standard and connector used for the external display connection.

DX	Long-distance, or rare, or wanted by you, amateur radio station. The abbreviation comes from the Morse telegraphy code for 'distant exchange.'
DXCC	The DX Century Club. An awards program based around confirming contacts with 100 DXCC 'countries' or 'entities' on various modes and bands. The DXCC list of 304 currently acceptable DXCC entities is used as the worldwide standard for what is a separate country or a recognizably separate island or geographic region.
DXpedition	A DXpedition is a single, or group of, amateur radio operators who travel to a rare or difficult to contact location, for the purpose of making contacts with as many amateur radio operators as possible worldwide. They often activate rare DXCC entities or islands, and they may operate stations on several bands and modes simultaneously.
EXT	Abbreviation for External
FFT	Fast Fourier Transformation – conversion of signals from the time domain to the frequency domain (and back).
FIX	Fixed mode. Sets the spectrum and waterfall display so that it displays the range frequencies between two pre-set frequencies determined a start frequency and the Span setting.
FM	Frequency Modulation. Used for repeaters on the 10m and 6m bands.
FPGA	Field Programmable Gate Array – a chip that can be programmed to act like logic circuits, memory, or a CPU.
FSK	Frequency Shift Keying.
FSK RTTY	FSK RTTY is keyed using a digital signal to offset the transmit frequency rather than AFSK which generates audio tones at the Mark and Space offsets from the VFO frequency.
GND GROUND	Ground. The earthing terminal for the radio. This should be connected to a 'telecommunications' ground spike, not the mains earth.
GPS	Global Positioning System. A network of satellites used for navigation, geolocation, and very accurate time signals.
GPSDO	GPS disciplined oscillator. An oscillator locked to time signals received from GPS satellites
Hex	Hexadecimal – a base 16 number system used as a convenient way to represent binary numbers. For example, 1001 1000 in binary is equal to 98h or 152 in decimal.
HF	High Frequency (3 MHz -30 MHz)
Hz	Hertz is a unit of frequency. 1 Hz = 1 cycle per second.

IF or I.F.	Intermediate frequency = the (Signal – LO) or (Signal + LO) output of a mixer. Signal = incoming signal frequency. LO = local oscillator frequency.
IMD	Intermodulation distortion. Interference/distortion caused by passing a wanted signal through a non-linear device such as a mixer or a power amplifier. There are IMD tests for receivers and transmitters. IMD performance of linear amplifiers can also be tested.
IPO	Intercept Point Optimization. Maximizes the dynamic range and enhances the close multi-signal and intermodulation characteristics of the receiver. Removes the AMP2 preamplifier and adds around 10 dB of attenuation to the signal.
IQ	Refers to the I and Q data streams treated as a pair of signals. For example, a digital signal carrying both the I (incident) and Q (quadrature) data.
Key	A straight key, paddle, or bug, used to send Morse Code
kHz	Kilohertz is a unit of frequency. 1 kHz = 1 thousand cycles per second.
LAN	Local Area Network. The Ethernet and WIFI connected devices connected to an ADSL or fiber router at your house is a LAN.
LED	Light Emitting Diode
LSB	Lower sideband SSB transmission
m, 6m	Meter (US) or Metre. Often used to denote an amateur radio or shortwave band; e.g. 6m, 20m, 10m, where it denotes the approximate free-space wavelength of the radio frequency. Wavelength = 300 / frequency in MHz. A frequency range of 3 to 30 MHz has a corresponding wavelength of 100m to 10m.
Marker	The receive (green) and transmit (red) frequency markers indicate the frequency of the VFO. The receiver marker is usually only visible when using split.
MCH	Memory channel
MDS	Minimum discernible signal. A measurement of receiver sensitivity
MHz	Megahertz – unit of frequency = 1 million cycles per second.
MIC	Microphone
MT	Memory Tune. If you select a memory channel and then turn the VFO knob to a different frequency the memory number above the mode indicator will change from the memory number to MT. The radio will behave as if it is in VFO mode.
NB	Noise Blanker. A filter used to eliminate impulse noise

Net	Usually refers to an on-air meeting of a group of amateur operators. Also, to 'Net' to the CW receive frequency means to adjust the transmitter or receiver frequency so that you will transmit CW on exactly the same frequency as you are receiving.
NR	Noise Reduction. A filter used to eliminate continuous background noise
Onboard	A feature performed within the radio. Especially one that usually requires external software. For example, the radio has 'onboard' CW, RTTY, and PSK decoders.
PC	Personal Computer. For the examples throughout this book, it means a computer running Windows 10.
Pileup	A pileup is a situation where a large number of stations are trying to work a single station. For example, a DXpedition or a rare DXCC entity. Split operation is often employed to spread the pileup of calling stations over a range of frequencies.
PO or Po	RF power output (meter)
PSK	Phase shift keying. A digital transmission mode that uses phase changes to transmit the bits in a data stream. The PSK decoder in the FTDX10 supports BPSK31 and QPSK31.
PROC	The Speech Processor, or compressor. Increases the average power of your transmission by decreasing the dynamic range of the audio signal. i.e. it makes the quiet parts louder.
PMS	Programmable memory scan. Scans between frequencies stored in memory slots M-P1L and M-P1U through M-P9L and M-P9U.
PTT	Press to talk - the transmit button on a microphone – The PTT signal sets the radio and/or software to transmit mode.
QMB	Quick memory button. A short-term memory function, storing either five or ten frequencies along with the mode and filter settings.
QPSK	Quadrature phase-shift keying. Digital transmission mode using 90-degree phase changes to indicate four two-bit binary states 00,01,10,11. The QPSK decoder in the FTDX10 supports QPSK31. QPSK is faster than BPSK31 but uses more bandwidth and is not as easy to decode. It usually has built-in error correction.
QRP	Q code - low power operation (usually less than 10 Watts).
QSO	Q code – an amateur radio conversation or "contact."
QSK	Q code – fast transmit to receive switching which allows Morse code to be received in the gaps between the CW characters that you are sending.
QSY	Q code – a request or decision to change to another frequency.

RBW	Resolution Bandwidth is the ability of the spectrum scope (spectrum and waterfall) to distinguish between signals that are on frequencies that are very close together. A high RBW can separately display signals that are closer together but requires more processing power.
RF	Radio Frequency
RS232	A computer interface used for serial data communications.
RTTY	Radio Teletype. RTTY is a frequency shift digital mode. Characters are sent using alternating Mark and Space tones.
RTTY-L	RTTY transmitted and received on lower sideband (NOR polarity)
RTTY-U	RTTY transmitted and received on upper sideband (NOR polarity)
RTS	'Ready to Send' is a com port control line often used for sending the PTT (SEND) command over the CAT interface between the radio and a PC.
RX	Abbreviation for receive or receiver
SDR	Software Defined Radio. The FTDX10 is more correctly a 'Hybrid SDR' as it does not use direct digital sampling except for the spectrum display.
Sked	A pre-organized or scheduled appointment to communicate with another amateur radio operator
SNR	Signal-to-Noise Ratio, usually stated in dB (decibels).
Soft Key	A button or selectable icon displayed on the touchscreen
Split	The practice of transmitting on a different frequency to the one that you are receiving on. Split operation is commonly used by DXpeditions and anyone who generates a large pileup of callers. SSB split is commonly 5-10 kHz. CW split is commonly 1-2 kHz.
Squelch	Squelch mutes the audio to the speakers when you are not receiving a wanted signal. When the received signal level increases the squelch opens and you can hear the station. The Squelch setting does affect the audio output over the USB cable.
SSB	Single Side Band transmission mode.
SWR	Standing Wave Ratio. A measurement of the ratio of the RF power reflected back from a mismatched antenna or connection, compared with the incident (transmitted) power. (Metered)
Tail	Furry attachment at the back of a dog or cat. Also, the length of time a repeater stays transmitting after the input signal has been lost, or a short flexible length of coaxial cable at the antenna or shack end of your main feeder cable.
TU	Morse code abbreviation meaning 'to you.' It is used when you have finished transmitting and wish the other station to respond.
TX	Abbreviation for Transmit or Transmitter

TXW	A button that lets you listen to the transmit frequency during Split operation.
UHF	Ultra High Frequency (300 MHz - 3000 MHz).
USB	Universal serial bus – serial data communications between a computer and other devices. USB 2.0 is fast. USB 3.0 is very fast.
USB	Upper sideband SSB transmission.
USOS	Unshift on space. Used to minimize transmission errors on RTTY.
VBW	Video Bandwidth is the ability of the spectrum scope (spectrum and waterfall) to distinguish weak signals from noise. A narrow VBW can filter noise but requires more processing power.
VFO	Variable Frequency Oscillator. The FTDX10 has two VFOs called 'VFO-A' and 'VFO-B' The main tuning knob controls the active VFO. In Split mode, if CS is turned off, the VFO tuning ring controls the alternate VFO (transmitter frequency).
VHF	Very High Frequency (30 MHz -300 MHz)
VOX	Voice Operated Switch. Voice-activated receive to transmit switching
W	Watts – unit of power (electrical or RF).
ZIN	This is a button which pulls the radio VFO onto the frequency that matches the CW Pitch that you have set. It 'Nets' the CW receive frequency to the transmitter frequency so that you will transmit CW on the same frequency as you are receiving. It only works in the CW mode.

Table of drawings and images

Index

The Author

Well, if you have managed to get this far you deserve a cup of tea and a chocolate biscuit. It is not easy digesting large chunks of technical information. It is probably better to dip into the book as a technical reference. Anyway, I hope you enjoyed it and that it has made life with the FTDX10 a little easier.

I live in Christchurch, New Zealand. I am married to Carol who is very understanding and tolerant of my obsession with amateur radio. She describes my efforts as "Andrew playing around with radios." We have two children and two cats. James has graduated from Canterbury University with a degree in Commerce. He is working for a large food wholesaler. Alex is a doctor working in the Wellington Hospitals.

I am a keen amateur radio operator who enjoys radio contesting, chasing DX, digital modes, and satellite operating. But I am rubbish at sending and receiving Morse code. I write extensively about many aspects of the amateur radio hobby. This is my tenth Amateur Radio book.

Thanks for reading my book!

73 de Andrew ZL3DW.

THE END
73 and GD DX

Quick Reference Guide

Function	FUNC menu setting
AGC settings (AM)	<FUNC> <RADIO SETTING> <MODE AM>
AGC settings (CW)	<FUNC> <CW SETTING> <MODE CW>
AGC settings (FM)	<FUNC> <RADIO SETTING> <MODE FM>
AGC settings (PSK/DATA)	<FUNC> <RADIO SETTING> <MODE PSK/DATA>
AGC settings (RTTY)	<FUNC> <RADIO SETTING> <MODE RTTY>
AGC settings (SSB)	<FUNC> <RADIO SETTING> <MODE SSB>
AM Mod from USB cable or DATA jack	<FUNC> <RADIO SETTING> <MODE AM> <REAR SELECT>
AM Mod level (Mic)	<FUNC> <RADIO SETTING> <MODE AM> <MIC GAIN>
AM Mod source (Mic or Rear)	<FUNC> <RADIO SETTING> <MODE AM> <AM MOD SOURCE>
AM OUT level to RTTY/DATA jack	<FUNC> <RADIO SETTING> <MODE AM> <AM OUT LEVEL>
AM Power - max RF power on AM	<FUNC> <OPERATION SETTING> <TX GENERAL> <AM MAX POWER>
AMC LEVEL	<FUNC> <AMC LEVEL>
APF filter frequency	CONT/APF knob
APF filter width	<FUNC> <OPERATION SETTING> <RX DSP> <APF WIDTH>
Audio level from RTTY/DATA port (AM)	<FUNC> <RADIO SETTING> <MODE AM> <RPORT GAIN>
Audio level from RTTY/DATA port (CW)	<FUNC> <RADIO SETTING> <MODE CW> <RPORT GAIN>
Audio level from RTTY/DATA port (FM)	<FUNC> <RADIO SETTING> <MODE FM> <RPORT GAIN>
Audio level from RTTY/DATA port (PSK/DATA)	<FUNC> <RADIO SETTING> <MODE PSK/DATA> <RPORT GAIN>
Audio level from RTTY/DATA port (SSB)	<FUNC> <RADIO SETTING> <MODE SSB> <RPORT GAIN>
Beep level	<FUNC> <OPERATION SETTING> <GENERAL> <BEEP LEVEL>
Beer	<FRIDGE> <OPEN BOTTLE> <DRINK CONTENTS>
BPSK / QPSK switch	<FUNC> <RADIO SETTING> <ENCDEC PSK> <PSK MODE>
Break-in delay	<FUNC> <BK-DELAY>
Break-in Semi or Full	<FUNC> <CW SETTING> <MODE CW> <CW BREAK-IN TYPE>
Break-in QSK delay	<FUNC> <CW SETTING> <MODE CW> <QSK DELAY TIME>

Callsign on splash screen during boot	<FUNC> <DISPLAY SETTING> <DISPLAY> <MY CALL>
Callsign on splash screen during boot delay	<FUNC> <DISPLAY SETTING> <DISPLAY> <MY CALL TIME>
CAT rate	<FUNC> <OPERATION SETTING> <GENERAL> <CAT RATE>
CAT RTS	<FUNC> <OPERATION SETTING> <GENERAL> <CAT RTS>
Compressor or speech processor level	<FUNC> <PROC LEVEL>
Contour filter depth or boost	<FUNC> <OPERATION SETTING> <RX DSP> <CONTOUR LEVEL>
Contour filter width	<FUNC> <OPERATION SETTING> <RX DSP> <CONTOUR WIDTH>
CS button functions	Press and hold CS button
CW APF filter frequency	CONT/APF knob
CW APF filter width	<FUNC> <OPERATION SETTING> <RX DSP> <APF WIDTH>
CW DECODE	<FUNC> <DECODE> - while in CW mode
CW decoder bandwidth	<FUNC> <CW SETTING> <DECODE CW> <CW DECODE BW>
CW Pitch offset or Filter display	<FUNC> <CW SETTING> <MODE CW> <CW FREQ DISPLAY>
CW indicator on filter display	<FUNC> <CW SETTING> <MODE CW> <CW INDICATOR>
CW initial contest number	<FUNC> <CW SETTING> <KEYER> <CONTEST NUMBER>
CW Messages	<FUNC> <MESSAGE> - while in CW mode
CW messages - auto repeat delay	<FUNC> <CW SETTING> <KEYER> <REPEAT INTERVAL>
CW messages - input via Text or Paddle	<FUNC> <CW SETTING> <KEYER> <CW MEMORY 1-5>
CW Number Style	<FUNC> <CW SETTING> <KEYER> <NUMBER STYLE>
CW OUT level to RTTY/DATA jack	<FUNC> <CW SETTING> <MODE CW> <CW OUT LEVEL>
CW sidetone level	in CW mode <FUNC> <MONI LEVEL> <turn FUNC>
CW wave shape	<FUNC> <CW SETTING> <MODE CW> <CW WAVE SHAPE>
CW while in SSB mode	<FUNC> <CW SETTING> <MODE CW> <CW AUTO MODE>
Date & Time setting	<FUNC> <EXTENSION SETTING> <DATE & TIME>
DNR level	Press and hold the DNR button
Emergency frequency enable	<FUNC> <OPERATION SETTING> <TX GENERAL> <EMERGENCY FREQ TX>

External display - enable	<FUNC> <DISPLAY SETTING> <EXT MONITOR> <EXT DISPLAY>
External display - 800x480 or 800x600	<FUNC> <DISPLAY SETTING> <EXT MONITOR> <PIXEL>
External tuner (INT, EXT, ATAS)	<FUNC> <OPERATION SETTING> <GENERAL> <TUNER SELECT>
Filter slope (AM)	<FUNC> <RADIO SETTING> <MODE AM>
Filter slope (CW)	<FUNC> <RADIO SETTING> <MODE CW>
Filter slope (FM)	<FUNC> <RADIO SETTING> <MODE FM>
Filter slope (PSK/DATA)	<FUNC> <RADIO SETTING> <MODE PSK/DATA>
Filter slope (RTTY)	<FUNC> <RADIO SETTING> <MODE RTTY>
Filter slope (SSB)	<FUNC> <RADIO SETTING> <MODE SSB>
Firmware - list current firmware versions	<FUNC> <EXTENSION SETTING> <SOFT VERSION>
Firmware update	<FUNC> <EXTENSION SETTING> <SD CARD> <Firmware Update DONE>
FM Mod from USB cable or DATA jack	<FUNC> <RADIO SETTING> <MODE FM> <REAR SELECT>
FM Mod source (Mic or Rear)	<FUNC> <RADIO SETTING> <MODE FM> <FM MOD SOURCE>
FM OUT level to RTTY/DATA jack	<FUNC> <RADIO SETTING> <MODE FM> <FM OUT LEVEL>
Keyboard language (external keyboard only)	<FUNC> <OPERATION SETTING> <GENERAL> <KEYBOARD LANGUAGE>
Keyed data shift (carrier point in DATA mode)	<FUNC> <RADIO SETTING> <MODE PSK/DATA> <DATA SHIFT (SSB)>
LED brightness	<FUNC> <DISPLAY SETTING> <DISPLAY> <LED DIMMER>
Manual notch filter frequency	NOTCH knob
Manual notch filter width	<FUNC> <OPERATION SETTING> <RX DSP> <IF NOTCH WIDTH>
Memory groups (on or off)	<FUNC> <OPERATION SETTING> <GENERAL> <MEM GROUP>
Mouse pointer speed	<FUNC> <DISPLAY SETTING> <DISPLAY> <MOUSE POINTER SPEED>
MPVD VFO tuning ring encoder steps per revolution	<FUNC> <OPERATION SETTING> <TUNING> <MPVD STEPS PER REV>
NB noise blanker level	Press and hold the NB button. Turn FUNC knob
NB noise blanker rejection (attenuation applied)	<FUNC> <OPERATION SETTING> <GENERAL> <NB REJECTION>
NB noise blanker width	<FUNC> <OPERATION SETTING> <GENERAL> <NB WIDTH>
Parametric equalizer (SSB transmitter)	<FUNC> <OPERATION SETTING> <TX AUDIO> <PRMTRC EQ...>

Parametric equalizer with speech processor turned on	<FUNC> <OPERATION SETTING> <TX AUDIO> <P PRMTRC EQ...>
PC keying line for CW	<FUNC> <CW SETTING> <MODE CW> <PC KEYING>
Power - max RF power on the HF bands	<FUNC> <OPERATION SETTING> <TX GENERAL> <HF MAX POWER>
Power - max RF power on the 6m band	<FUNC> <OPERATION SETTING> <TX GENERAL> <50M MAX POWER>
Power - max RF power on the 70 MHz band	<FUNC> <OPERATION SETTING> <TX GENERAL> <70M MAX POWER>
Power - max RF power on AM	<FUNC> <OPERATION SETTING> <TX GENERAL> <AM MAX POWER>
PSK DECODE	<FUNC> <DECODE> - while in PSK mode
PSK Messages	<FUNC> <MESSAGE> - while in PSK mode
PSK tone offset	<FUNC> <RADIO SETTING> <MODE PSK/DATA> <PSK TONE>
PSK transmit level from the RTTY/DATA jack	<FUNC> <RADIO SETTING> <ENCDEC PSK> <PSK TX LEVEL>
PSK/DATA Mod from USB cable or DATA jack	<FUNC> <RADIO SETTING> <MODE PSK/DATA> <REAR SELECT>
PSK/DATA mod source (Mic or Rear)	<FUNC> <RADIO SETTING> <MODE PSK/DATA> <PSK/DATA MOD SOURCE>
PSK/DATA OUT level to the RTTY/DATA jack	<FUNC> <RADIO SETTING> <MODE PSK/DATA> <DATA OUT LEVEL>
QMB 5 or 10 memory slots	<FUNC> <OPERATION SETTING> <GENERAL> <QMB CH>
QPSK Polarity Settings	<FUNC> <RADIO SETTING> <ENCDEC PSK> <QPSK POLARITY (RX or TX)>
QSK delay	<FUNC> <CW SETTING> <MODE CW> <QSK DELAY TIME>
Quick split input popup enable	<FUNC> <OPERATION SETTING> <GENERAL> <QUICK SPLIT INPUT>
Quick split offset	<FUNC> <OPERATION SETTING> <GENERAL> <QUICK SPLIT FREQ>
Recall FUNC menu settings from SD card	<FUNC> <EXTENSION SETTING> <SD CARD> <Menu Load DONE> <Select a file>
Recall memory channels from SD card	<FUNC> <EXTENSION SETTING> <SD CARD> <Mem List Load DONE> <Select a file>
Repeater offset 10m band	<FUNC> <RADIO SETTING> <MODE FM> <RPT SHIFT (28MHz)>
Repeater offset 6m band	<FUNC> <RADIO SETTING> <MODE FM> <RPT SHIFT (50MHz)>
Reset - Full reset	<FUNC> <EXTENSION SETTING> <RESET> <ALL RESET>
Reset - Memory slots	<FUNC> <EXTENSION SETTING> <RESET> <MEMORY CLEAR>

Reset - Menu settings	<FUNC> <EXTENSION SETTING> <RESET> <MENU CLEAR>
RF Gain or Squelch	<FUNC> <OPERATION SETTING> <GENERAL> <RF/SQL VR>
RF Power setting	<FUNC> <RF POWER> <turn MULTI>
RS232 port rate & timeout	<FUNC> <OPERATION SETTING> <GENERAL> <232C RATE> or <232C TIME OUT TIMER>
RTS/DTR/DAKY selection (AM)	<FUNC> <RADIO SETTING> <MODE AM> <RPPT SELECT>
RTS/DTR/DAKY selection (FM)	<FUNC> <RADIO SETTING> <MODE FM> <RPPT SELECT>
RTS/DTR/DAKY selection (PSK/DATA)	<FUNC> <RADIO SETTING> <MODE PSK/DATA> <RPPT SELECT>
RTS/DTR/DAKY selection (RTTY)	<FUNC> <RADIO SETTING> <MODE RTTY> <RPPT SELECT>
RTS/DTR/DAKY selection (SSB)	<FUNC> <RADIO SETTING> <MODE SSB> <RPPT SELECT>
RTTY DECODE	<FUNC> <DECODE> - while in RTTY mode
RTTY Mark frequency	<FUNC> <RADIO SETTING> <MODE RTTY> <MARK FREQUENCY>
RTTY Messages	<FUNC> <MESSAGE> - while in RTTY mode
RTTY OUT level to RTTY/DATA jack	<FUNC> <RADIO SETTING> <MODE RTTY> <RTTY OUT LEVEL>
RTTY Polarity (RX)	<FUNC> <RADIO SETTING> <MODE RTTY> <POLARITY RX>
RTTY Polarity (TX)	<FUNC> <RADIO SETTING> <MODE RTTY> <POLARITY TX>
RTTY Shift	<FUNC> <RADIO SETTING> <MODE RTTY> <SHIFT FREQUENCY>
RTTY text and Baudot code settings	<FUNC> <RADIO SETTING> <ENCDEC RTTY>
Save FUNC menu settings to SD card	<FUNC> <EXTENSION SETTING> <SD CARD> <Menu Save DONE> <NEW> <ENT>
Save memory channels to SD card	<FUNC> <EXTENSION SETTING> <SD CARD> <Mem List Save DONE> <NEW> <ENT>
Scan - Mic scan enable	<FUNC> <OPERATION SETTING> <GENERAL> <MIC SCAN>
Scan - Mic scan resume time	<FUNC> <OPERATION SETTING> <GENERAL> <MIC SCAN RESUME>
Screensaver start delay	<FUNC> <DISPLAY SETTING> <DISPLAY> <SCREEN SAVER>
SD Card - capacity and free space information	<FUNC> <EXTENSION SETTING> <SD CARD> <Informations DONE>
SD Card - format	<FUNC> <EXTENSION SETTING> <SD CARD> <Format DONE> <DONE>

SD card - save FUNC menu settings	\<FUNC\> \<EXTENSION SETTING\> \<SD CARD\> \<Menu Save DONE\> \<NEW\> \<ENT\>
SD card - recall FUNC menu settings	\<FUNC\> \<EXTENSION SETTING\> \<SD CARD\> \<Menu Load DONE\> \<Select a file\>
SD Card - save memory channels	\<FUNC\> \<EXTENSION SETTING\> \<SD CARD\> \<Mem List Save DONE\> \<NEW\> \<ENT\>
SD Card - recall memory channels	\<FUNC\> \<EXTENSION SETTING\> \<SD CARD\> \<Mem List Load DONE\> \<Select a file\>
Spectrum scope 2D display sensitivity	\<FUNC\> \<DISPLAY SETTING\> \<SCOPE\> \<2D DISP SENSITIVITY\>
Spectrum scope 3DSS display sensitivity	\<FUNC\> \<DISPLAY SETTING\> \<SCOPE\> \<3DSS DISP SENSITIVITY\>
Spectrum scope, filter center or carrier point display	\<FUNC\> \<DISPLAY SETTING\> \<SCOPE\> \<SCOPE CTR\>
Spectrum scope resolution bandwidth	\<FUNC\> \<DISPLAY SETTING\> \<SCOPE\> \<RBW\>
Speech processor level	\<FUNC\> \<PROC LEVEL\>
Split - Quick split input popup enable	\<FUNC\> \<OPERATION SETTING\> \<GENERAL\> \<QUICK SPLIT INPUT\>
Split - Quick split offset	\<FUNC\> \<OPERATION SETTING\> \<GENERAL\> \<QUICK SPLIT FREQ\>
Squelch or RF Gain control	\<FUNC\> \<OPERATION SETTING\> \<GENERAL\> \<RF/SQL VR\>
SSB Mod from USB cable or DATA jack	\<FUNC\> \<RADIO SETTING\> \<MODE SSB\> \<REAR SELECT\>
SSB Mod source (Mic or Rear)	\<FUNC\> \<RADIO SETTING\> \<MODE SSB\> \<SSB MOD SOURCE\>
SSB OUT level to RTTY/DATA jack	\<FUNC\> \<RADIO SETTING\> \<MODE SSB\> \<SSB OUT LEVEL\>
Time & Date setting	\<FUNC\> \<EXTENSION SETTING\> \<DATE & TIME\>
Touchscreen brightness	\<FUNC\> \<DIMMER\>
Touchscreen calibration	\<FUNC\> \<EXTENSION SETTING\> \<CALIBRATION\>
Touchscreen contrast	\<FUNC\> \<CONTRAST\>
Transmitter bandwidth (AM)	\<FUNC\> \<RADIO SETTING\> \<MODE AM\> \<TX BPF SEL\>
Transmitter bandwidth (FM)	\<FUNC\> \<RADIO SETTING\> \<MODE FM\> \<TX BPF SEL\>
Transmitter bandwidth (PSK/DATA)	\<FUNC\> \<RADIO SETTING\> \<MODE PSK/DATA\> \<TX BPF SEL\>
Transmitter bandwidth (SSB)	\<FUNC\> \<RADIO SETTING\> \<MODE SSB\> \<TX BPF SEL\>
Transmit monitor level	in a voice mode \<FUNC\> \<MONI LEVEL\> \<turn FUNC\>
Transmit time-out timer	\<FUNC\> \<OPERATION SETTING\> \<GENERAL\> \<TX TIME OUT TIMER\>

Tuning step dial rate AM	<FUNC> <OPERATION SETTING> <TUNING> <AM CH STEP>
Tuning step dial rate FM	<FUNC> <OPERATION SETTING> <TUNING> <FM CH STEP>
Tuning step dial rate SSB, CW, and other modes	<FUNC> <OPERATION SETTING> <TUNING> <CH STEP>
VFO encoder steps per revolution	<FUNC> <OPERATION SETTING> <TUNING> <MAIN STEPS PER REV>
VFO MPVD tuning ring encoder steps per revolution	<FUNC> <OPERATION SETTING> <TUNING> <MPVD STEPS PER REV>
VFO step size RTTY & PSK 5 Hz or 10 Hz	<FUNC> <OPERATION SETTING> <TUNING> <RTTY/PSK DIAL STEP>
VFO step size SSB & CW 5 Hz or 10 Hz	<FUNC> <OPERATION SETTING> <TUNING> <SSB/CW DIAL STEP>
Voice Messages	<FUNC> <MESSAGE> - while in SSB, AM, or FM mode
VOX level from DATA jack	<FUNC> <OPERATION SETTING> <TX GENERAL> <DATA VOX GAIN>

Figure 56: Size comparison. The FTDX10 on top of a FTDX101D

Made in United States
Troutdale, OR
08/19/2024

22136191R00115